Ab   Service

1

T

THE EDGE OF SCIENCE

The Edge of
Science

Mysteries of Mind, Space and Time

ALAN BAKER

MAINSTREAM
PUBLISHING

EDINBURGH AND LONDON

First published in Great Britain in 2009 by
MAINSTREAM PUBLISHING COMPANY
(EDINBURGH) LTD
7 Albany Street
Edinburgh EH1 3UG

ISBN 9781845963378

A catalogue record for this book is
available from the British Library

Typeset in OptimusPrinceps and Sabon

Printed in Great Britain by
CPI Mackays of Chatham Ltd, Chatham, ME5 8TD

CONTENTS

INTRODUCTION

The history of human civilisation is the history of the search for answers. In all likelihood, the very first questions asked by human beings were not consciously expressed: they were more likely merely instinctive imperatives based on the need to survive, to avoid predators and find enough food to eat. In the modern world, of course, those imperatives are still of supreme importance (although now the 'predators' are much more likely to be other humans than hungry wild animals); but such is the success of human civilisation that we can ask many other questions based on our observations of the world, and formulate theories which point towards possible answers.

For instance, over the centuries philosophers, scientists and ordinary men and women have asked what is perhaps the ultimate question: how did the Universe begin? Did it emerge from an unthinkably violent primordial fireball, an event which the great astronomer Sir Fred Hoyle [1915–2001] derisively called the 'Big Bang'? Or did it begin in another way?

Many people will have heard of the mysterious 'Face on Mars', the kilometre-long mesa which was revealed in one of the photographs taken by the US Viking probe in the late 1970s. Eerily reminiscent of a human face (at least in that first photograph), further higher-definition imaging by later probes has revealed it to be merely an oddly shaped lump of rock, with little or no resemblance to a face, human or otherwise.

And yet there are many other curious and intriguing features on the surface of the red planet, including pyramid-like structures and patterns that appear to be the products of intelligent design. The Moon, also, contains some remarkable features, including the so-called 'Shard', a skyscraper-like object rising several kilometres above the lunar surface, and the 'Castle', a metallic agglomeration of rods and domes which appears to be hanging from a thin filament high above the surface. Are these bizarre objects on the Moon and Mars evidence of an ancient alien presence on our planetary neighbours? Orthodox astronomers and planetary scientists are unwilling seriously to entertain this notion; but are they wrong?

Allied to this question is that of extraterrestrial life and intelligence. The extraterrestrial alien is one of the great cultural icons of the twentieth century. It is permanently rooted in the human mind and imagination – and with good reason: with at least 100 billion stars in our own galaxy alone, and at least 100 billion galaxies in the observable Universe, the chances of there *not* being alien life are vanishingly small. From the ice-covered ocean of Jupiter's moon Europa, to the planets we now know are orbiting other stars, the search for extraterrestrial life (and perhaps intelligence) is gathering pace. Will we ever find it?

This book examines these and many other questions that have bewildered and perplexed humanity for centuries. Some are being addressed by scientists on a daily basis, while others are considered too bizarre to be embraced by the mainstream (although that does not mean that they are unworthy of serious attention and investigation).

There are other mysteries which many people don't even *realise* are mysteries. Take the Moon, for instance. Mottled with pale blue-grey 'seas' surrounded by bright highlands, the Moon is perhaps the most beautiful sight to be seen in the skies of Earth. It is the closest celestial body to our world, and its unusually large size compared with its 'primary' has maintained the stability of Earth's axis and rotation, allowing life and intelligence to develop. We would not exist without the Moon. And yet the Moon's origin is still a puzzle to astronomers. There are a number of theories to account for its existence: that it was a rogue planet from elsewhere in the universe which was captured aeons ago by Earth's gravitational field; that it formed out of the same cloud of dust and gas as Earth and all the other planets of the Solar System; that it is the result of a titanic collision between the early Earth and a Mars-sized planet while the Solar System was still forming. However, there are problems with each of these theories. What is the true origin of the Moon?

We like to think that the Earth's surface has been pretty much explored, and that we are the masters thereof. Of course, there are still a few blank spaces on the map, but they are shrinking rapidly as humanity continues to exercise its complete (and frequently ill-considered) dominion over the planet. And yet there are still strange and terrifying mysteries out there, and not always in the less-travelled regions of our world. Strange creatures such as the Mothman haunt our cities and strike terror into witnesses. Orthodox science refuses to accept that they exist, but it seems that orthodox science is wrong ...

These are just a handful of the mysteries of science that we will examine in this book. It will be a strange and sometimes frightening journey, ranging from the moment of the Universe's creation to its final moments countless aeons in the future. But I hope that this journey to the 'edge of science' will be one that the reader will find intriguing and stimulating.

Acknowledgment is gratefully made for permission to quote material from the following published works:

From *Arktos: The Polar Myth in Science, Symbolism and Nazi Survival* by Joscelyn Godwin, © 1993. Reprinted by kind permission of Thames & Hudson Ltd. London.

From *The Physics of Immortality* ©1994 by Frank J. Tipler, Pan Books, London.

From *Fads and Fallacies in the Name of Science* © 1952 by Martin Gardner, Dover Publications, New York.

1

ORIGIN

HOW DID THE UNIVERSE BEGIN?

It is one of the setbacks of living in a city that so few stars are visible in the night sky. The light pollution from buildings and streetlamps washes the darkness of the great dome of space above us with a depressing sodium glow, almost completely obscuring the majestic sweep of stars that stretch into apparent infinity around our tiny world. To the more poetically-minded, this may be a cause of sadness; it is as if the trappings of technological society have, ironically, drawn a veil across the face of the cosmos, divorcing us from the unthinkably magnificent Universe which gave birth to us and to everything we know. Astronomers constantly bemoan the presence of light pollution, and with good reason, since it robs city dwellers of the opportunity to partake of the wonder of the stars and galaxies that surround us.

Anyone who has been lucky enough to stand in a desert or other wilderness, far from human habitation and the sprawl of cities, and who has looked up into the night sky, will be staggered by the profusion of stars splashed across the firmament. It is a dizzying experience, and one which is not easily forgotten. Indeed, the same poetic types who dwell in the sickly sodium-orange glow of cities may be momentarily seized with fear when looking up at a sky uncluttered by terrestrial light. One has the eerie and vertiginous feeling that there is *nothing* – no protective atmosphere, no warm air to breathe – between oneself and the limitless immensity of the

Universe beyond; nothing, that is, save vast and unimaginable distances.

At times like this, one may well wonder how all this came to be: ruddy Mars and bright Venus and the other worlds of the Solar System; the tiny pinpoint of Proxima Centauri (the nearest star to the Sun); the other stars of the local group; the great stellar sweep extending beyond to the fringes of our galaxy, the Milky Way; and the hundreds of millions of other galaxies wheeling through the eternal darkness of deep space, out to the very edge of the observable Universe. Where did all this come from? How and why did the Universe come into existence?

This is perhaps the oldest of all the many questions humanity has asked. How close are we to an answer?

THE BIG BANG

It was the Belgian astrophysicist and priest George Édouard Lemaître (1894–1966) who first suggested that the Universe emerged out of a single 'primeval atom' or 'cosmic egg'. After completing a PhD in physics at the University of Louvain in 1920, and further studies at Cambridge, Harvard and MIT, Lemaître developed the idea that the Universe had originated in a sphere no more than 30 times bigger than the Sun, which had exploded between 20 billion and 60 billion years ago. Although he was mistaken in some respects, Lemaître's basic concept formed the foundation upon which later theories were built.

At about the same time, the great American astronomer Edwin Hubble (1889–1953) discovered that the countless billions of galaxies of which the Universe is composed are not floating in stillness in the void of space, but are actually hurtling away from each other at tremendous speed – in other words, that the Universe is expanding.

Hubble was able to demonstrate this with the aid of the enormous 100-inch Hooker telescope at the Mount Wilson Observatory in California, through which he studied the light from distant galaxies. It is one of the wonders of the Universe that a beam of light can tell us so much about the place it came from. All that is needed to unlock its secrets is a prism, which splits the light passing through it into the spectrum of colours of which it is composed. The light coming from a distant source, such as a galaxy, contains various patterns caused by the colours of the light emitted by the different atoms which make up that source.

During the 1920s and '30s, while analysing the light from distant galaxies, Hubble noticed something strange and intriguing: the patterns had longer wavelengths than those coming from nearby stars; they were *stretched* towards the red end of the spectrum. The more distant the galaxy, the greater this 'redshift' became; and Hubble calculated that the redshift of a galaxy was proportional to its distance from Earth. In other words, the further away a galaxy was, the faster it was receding from us.

The inevitable conclusion of the 'Hubble law' is that, if the Universe is expanding, it must have had a single point of origin in the distant past, between 10 and 20 billion years ago, when all the matter and energy we now observe was compressed into an infinitely small, infinitely dense region.

This idea was unpalatable to some, including the great British astronomer Fred Hoyle, who disliked the notion of the Universe being created in a massive (and unexplained) primordial explosion. In fact, so unimpressed was he that he derisively dubbed this explosion the 'Big Bang', adding that it was about 'as elegant as a party girl jumping out of a cake'.

In the 1940s Hoyle proposed an alternative model for the Universe, together with his Cambridge University colleagues Hermann Bondi and Thomas Gold, which came to be known as the Steady State model. According to this hypothesis, which was partly inspired by a film called *The Dead of Night*, in which the last of four interlocked ghost stories ends at the beginning of the first story, the Universe does not have a beginning: it has always existed, and the expansion is due to the constant creation of new matter at certain locations such as the centres of galaxies.

Hoyle also hypothesised the existence of the 'C-field' (or 'creation field'), which pervades the entire Universe, but which is at its most powerful in the regions where matter is continuously being created. Hoyle suggested that the C-field causes the Universe to expand, counteracting the opposing tendency to contract as a result of the gravity of the Universe's contents. The C-field is carried by fundamental particles called bosons, which gain energy as they fall into regions of high gravity, creating new particles as they do so. As the C-field is strengthened by this process, an outward explosion occurs, similar to the Big Bang.

However, in his book *Companion to the Cosmos*, the British astrophysicist John Gribbin notes that the C-field hypothesis does not allow the formation of black holes or the creation of singularities (a singularity is a region of infinitely small size and density, where the laws of physics as we know them break down). According to Gribbin: 'In any situation where gravity tries to compress matter to such extremes, the C-field will reverse the process in an outward blast of new particles.'

The very presence of black holes in our own galaxy and beyond is a forceful argument against the C-field (although, as we shall see, the C-field's 'anti-gravitational' properties have resurfaced in recent years in the form of so-called 'dark energy' – see Chapter 12). In addition, an accidental discovery made in 1965 was to cast further doubt on the Steady State model put forward by Hoyle, Bondi and Gold.

This discovery, which was made by Arno Penzias and Robert Wilson, and which Gribbin calls 'the most important observation made in cosmology since the discovery by Edwin Hubble that the Universe is expanding', occurred at the Bell Research Laboratories in Holmdel, New Jersey. Penzias and Wilson had been working with a 20-foot horn antenna that had been designed for use with the Echo communications satellites, which in turn were being used to investigate radio emissions from the Milky Way.

One of the main problems in their work was the necessity of eliminating radio noise from terrestrial sources, which interfered with the signals they were attempting to detect. One particular source of interference remained persistent: a background of microwave radio noise which seemed to come from every direction in the sky.

In astronomy, as in most other things, timing is everything; and it just so happened that while Penzias and Wilson were puzzling over the strange, uniform microwave radio noise they were detecting with their antenna, another group were preparing to begin observations with a small radio telescope at Princeton University, to search for the very same radiation (which had been hypothesised as far back as the 1940s). When

Arno Penzias called the Princeton group's leader, Robert Dicke, to ask him if he had any idea what might be causing the ubiquitous radio noise, the answer became obvious. As Gribbin notes, 'the theory and the observations had come together, and two and two were quickly put together to make four'. In 1978 Penzias and Wilson were awarded the Nobel Prize for their discovery of the cosmic background radiation, the echo of the Big Bang, sometimes known as 'the afterglow of Creation'.

A few years later, however, astronomers began to grow uneasy with the smoothness and regularity of the cosmic background radiation, which came from every direction at exactly the same temperature of 2.7 degrees K. ('K' refers to the Kelvin temperature scale, which corresponds to Celsius, except that absolute zero, or $-273.16°C$, is $0°K$.) The problem was that if the Universe was perfectly smooth following the Big Bang, there seemed to be no way in which atoms could come together to form stars, planets and galaxies. In order for the Universe to develop into what we see today (and thus in order for us to exist to do the seeing!), there had to be 'ripples' in the background radiation, or slight variations in the temperature depending on which region of the sky one is looking at.

It was not until the launch of NASA's COBE satellite in 1992 that the mystery was solved. COBE (short for Cosmic Background Explorer) detected the ripples which astronomers had been searching for, offering further support for the Big Bang model of the Universe's origin.

If you require a demonstration of cosmic background radiation, simply detune your TV set, so that the antenna is not picking up any station. About 1 per cent of the 'snow' on your screen is caused by the background radiation left over from the Big Bang approximately 15 billion years ago.

MEMBRANE THEORY

According to Stoic philosophy, the Universe is periodically consumed by fire and regenerated in a process known as *ekpyrosis* (a Greek word meaning 'conflagration'). And *ekpyrosis* is the term applied to a brand new theory of the creation of the Universe, which University of Chicago cosmologist Michael Turner describes as being 'almost crazy enough to be correct'.

The theory, called the Ekpyrotic model, does not replace the Big Bang model, but it does do away with one of the Big Bang's most fundamental elements, *inflation*, which suggests that in the first split second of its existence, the Universe underwent a colossal (and colossally brief) period of accelerated expansion.

In his *Companion to the Cosmos*, John Gribbin writes:

> Taken at face value, the observed expansion of the Universe implies that it was born out of a singularity, a point of infinite density, 15–20 billion years ago. Quantum physics tells us that it is meaningless to talk in quite such extreme terms, and that instead we should consider the expansion as having started from a region no bigger across than the Planck length (10^{-35} m), when the density was not infinite but 'only' some 10^{94} grams per cubic centimetre. The first puzzle is how anything that dense could ever expand – it would have an enormously strong gravitational field, turning it into a black hole and snuffing it out of existence (back into the singularity) as soon as it was born . . .

The inflationary model was developed to resolve this problem by invoking a force which gave the primordial Universe a sudden violent outward 'push', in effect to cause its accelerated expansion before gravity could overcome it and cause it to fall instantly back into the singularity.

The new theory presented in May 1995, put forward by Neil Turok of Cambridge University, Burt Ovrut of the University of Pennsylvania, and Paul Steinhardt and Justin Khoury of Princeton University, suggests that 'our current Universe is a four-dimensional membrane embedded in a five-dimensional "bulk" space, something like a sheet of paper in ordinary three-dimensional space'.

According to this theory, there is a boundless fifth dimension, 'bulk space', in which are embedded 'membranes' (or 'branes' for short) containing the four dimensions of spacetime (three space dimensions and one time dimension). The four creators of this theory provided the website Space.com with a short essay in which they describe its basic features.

The Ekpyrotic Universe does not require inflation to smooth out the irregularities in its early structure, since the Big Bang is caused by a collision between two branes floating in bulk space. According to Turok *et al.*, when two three-dimensional branes collide and 'stick', the kinetic energy of the collision is converted to the fundamental particles of which the Universe is composed, and which are confined to movement within the three space dimensions of the brane. In other words, there is no need for the troublesome singularity of the standard Big Bang model.

The 'ripples' in the cosmic background radiation, which were detected by the COBE satellite mentioned earlier, are caused by quantum effects in the 'incoming three-dimensional world . . . so that the collision occurs in some places at slightly different times than others. By the time the collision is complete, the rippling leads to small variations in temperature, which seed temperature fluctuations in the microwave background and the formation of galaxies.'

Turok and his colleagues conclude with the caveat that their theory is brand new, while the inflationary model has been intensively studied for more than 20 years. For this reason, they caution against accepting the Ekpyrotic model too quickly: there is a great deal of work to be done before we can say that this may well be the way our Universe came into existence.

2

ARE WE THE
EXTRATERRESTRIALS?

PANSPERMIA AND THE ORIGIN OF LIFE

COSMIC ANCESTRY

One of the great unanswered questions of human history is: where did we come from? How did life on Earth originate? What started the great sequence of chemical and biological events whereby the first self-replicating molecules began to develop ever-greater complexity, culminating in the vast biodiversity we see around us today?

Since the 1920s, when the Russian biochemist Alexander Oparin and the English geneticist J.B.S. Haldane suggested independently that the generation of life could occur spontaneously from inanimate matter, the 'Oparin-Haldane paradigm' has been the favoured theory of the origin of life, with life originating in what is often (and rather inelegantly) called the 'primordial soup' of primitive chemical compounds existing on the early Earth. Over countless millennia, ultraviolet radiation from the Sun combined with powerful lightning strikes to produce more complex compounds until eventually, by the merest chance, a compound was created that was capable of self-replication.

There seemed to be no way to test this theory, until the 1950s, when Stanley Miller, a PhD candidate at the University of Chicago, designed an experiment which he believed would closely approximate the conditions on the primordial Earth at the time of life's beginnings. By combining methane and ammonia in a flask with a small amount of water and applying electricity, Miller succeeded in proving that amino acids and other organic molecules, the basic building blocks of life, could spontaneously arise in the right conditions. In Miller's miniature (and vastly simplified) version of the early Earth, the methane and ammonia acted as the atmosphere, and the electricity represented the primordial lightning.

There is, however, another theory on how life began on Earth, which is steadily gaining ground as evidence in its favour mounts. That theory is known as 'panspermia' (literally, 'seeds everywhere'), and it suggests that the building blocks of life may have come to Earth from the depths of interstellar space.

The first-known mention of the idea is in the writings of the fifth-century BC Greek philosopher Anaxagoras, although for 2,000 years it remained dormant, like the space seeds themselves, until it was revived in the late nineteenth and early twentieth centuries by scientists such as Hermann von Helmholtz and Svante Arrhenius. Strictly speaking, the correct term for this hypothesis is 'exogenesis', which states simply that life (or the material necessary for life) was transferred to Earth from elsewhere in the Universe, while saying nothing about how widespread that material actually is. Panspermia, on the other hand (as its name suggests), is the hypothesis that the material necessary to the establishment of life exists throughout the entire cosmos, and probably led to the development of life on many other planets besides Earth.

Perhaps the most important proponents of panspermia in recent years were the British astronomers Fred Hoyle and Chandra Wickramasinghe (b. 1939), whose spectroscopic observations of light from distant stars revealed organic compounds in the intervening interstellar dust. Hoyle and Wickramasinghe also suggested that comets (which are composed of dust and water ice) might carry bacterial life across the immensity of space, protecting it from the damaging effects of interplanetary and interstellar radiation. Even more controversially, they proposed that extraterrestrial biological material is *still* raining down upon the Earth from the depths of the Universe, and may be the cause of the viral epidemics that have periodically proven so deadly in human and animal populations.

The two astronomers later expanded their theory to take account of the evolutionary process and the vast complexity and variety of terrestrial life, which, they contended, could not be accounted for merely by random mutation – even over the roughly four billion years since life first arose.

DIRECTED PANSPERMIA

In 1973 the Nobel Prize-winning biochemist Francis Crick (co-discoverer, with James Watson, of the molecular structure of DNA) and Dr Leslie Orgel proposed the theory of 'directed panspermia', which suggests that the seeds of life may have been intentionally sown on Earth (and perhaps other planets in the Galaxy) by a highly advanced extraterrestrial civilisation.

One can easily imagine such a civilisation, existing somewhere in the limitless night of space, which perhaps faced an imminent catastrophe, or which perhaps wished to seed distant and yet-to-be-discovered worlds with living material from its own biosphere, so that one day those worlds might be colonised by later generations. Crick and Orgel suggested that it would be a relatively simple matter to fire small grains containing the building blocks of life in all directions into space. After countless aeons spent drifting through the void, some of those grains might well come to rest on the surfaces of young planets, where those building blocks might then commence their long evolutionary journey towards diversity and complexity.

If directed panspermia seems at first glance to be an unlikely scenario, more fitted to science fiction than to serious theorising on the origin of life on Earth, it is worth considering the worries that NASA and other space agencies have about the possibility of our *own* space exploration resulting in other worlds being contaminated by terrestrial micro-organisms and other organic materials. One only has to look at the 'clean rooms' in which our interplanetary probes are constructed to understand that such concerns are very real. Not only do mission planners have to ensure that planets such as Mars are not contaminated by our spacecraft, but they also must ensure that in our ongoing search for extraterrestrial life, we do not inadvertently mistake our own microbes for alien ones!

THE EVIDENCE FOR PANSPERMIA

The concept of panspermia is a bit like the science of 'exobiology', which attempts to form a picture of what extraterrestrial life forms might be like, and which has been called 'a science without a subject'. As hypotheses, panspermia and exogenesis are extremely difficult to test; and yet there is intriguing (if circumstantial) evidence in their favour. For instance, in 1996 astonishing claims were made regarding a meteorite known as ALH84001, and were reported by news organisations across the globe. This meteorite originally came from Mars, and was most likely ejected into space following a large meteorite impact on the red planet in the distant past. For uncounted years the fragment of Martian rock drifted through space until it finally entered the Earth's atmosphere and impacted in Antarctica.

When examined under a microscope, ALH84001 was found to contain strange structures reminiscent of terrestrial microfossils, which were quickly heralded as the first concrete evidence of extraterrestrial life. So astounding was this apparent discovery that the then US President, Bill Clinton, made an official televised announcement. However, scientific opinion is still divided on whether the tiny worm-like structures within the meteorite are really fossilised Martian life forms, or whether they were formed 'abiotically' and merely *resemble* once-living material.

If Fred Hoyle and Chandra Wickramasinghe's panspermia hypothesis is correct (if, in other words, extraterrestrial biological material is still coming to Earth from deep space), one might expect it to be possible to detect it directly, and even to take samples of it. Intriguingly, Wickramasinghe claims to have done just that. In November 2000, he reported that the Indian Space Research Organisation had collected micro-organisms from the Earth's atmosphere at an altitude of 16 km, and that they bore no resemblance to anything known on Earth. Less than a year later, in April 2001, scientists from India and the United Kingdom claimed to have retrieved extraterrestrial microbial life forms from even higher in the Earth's stratosphere, at an altitude of 41 km. They presented their findings at the 46th annual meeting of the International Society for Optical Engineering in San Diego, California. They added that 41 km is too high for air to rise naturally from lower regions, unless it was cast upwards in a powerful volcanic eruption. Even if this was the case, and the organisms had a terrestrial origin, it still goes some way to validating the panspermia hypothesis: in other words, either the organisms are of extraterrestrial origin, or it is possible for terrestrial organisms to be cast into near space by geological processes.

One of the principal objections to panspermia is that it would be extremely difficult – indeed impossible – for bacteria and other microscopic life forms to survive the intense rigours of interplanetary or interstellar travel. The combination of hard vacuum, intense cold and stellar radiation found in deep space would surely destroy them long before they had a chance to complete their billion-year journeys and find sanctuary in hospitable environments such as planetary surfaces.

This may not be the case, however. Right here on Earth we have organisms known as 'extremophiles': bacteria that can survive and even thrive in environments which should be lethal to them. Extremophile bacteria have been discovered at the bottoms of oceans around volcanic vents where the water is superheated to above 100°C, and even several kilometres beneath the ocean floor, where heat and pressure should prevent anything from surviving. Other bacteria are as comfortable in extreme cold as extreme heat, and have been discovered in a dormant state in ice cores taken from more than 1.6 km beneath the Antarctic ice pack. There is even a bacterium, known as *Deinoccocus radiodurans*, which is perfectly at home on the cooling rods of nuclear reactors! If such bacteria can survive, whether in active or dormant states, in such harsh conditions here on Earth, might it be possible for other life forms to endure the long, airless night of deep space, drifting between worlds like celestial pollen?

ARTIFICIAL CELLS

The panspermia hypothesis gained an additional boost in 2000 with the work of NASA researchers from the Ames Astrochemistry Laboratory and the Department of Chemistry and Biochemistry at the University of California, Santa Cruz. In an experiment reminiscent of Stanley Miller's groundbreaking work in the 1950s, the researchers bombarded simple, common chemicals with ultraviolet radiation. The chemicals, which were held in a vacuum to simulate the conditions in deep space, were in the form of ices composed of water, methanol, ammonia and carbon monoxide.

The result was the formation of solid materials which, when immersed in water, created membranous structures like soap bubbles. This is important because membranes (structures that protect the chemistry of life from the external environment) are essential to life's development. In addition, the NASA experiments suggest that if such 'cells' exist in comets, asteroids or interplanetary dust, then the early chemical stages in the origin of life do not require a planet. According to the science journalist Dr David Whitehouse: 'This implies that the vastness of space is filled with chemical compounds which, if they land in a hospitable environment like our Earth, can readily jump-start life.'

In an interview with the BBC, one of the researchers, Dr Scott Sandford, stated that the team would continue to investigate their discovery. 'We want to see if [the "cells"] show some of the same behaviours you see in biological membranes. It's one thing to have little structures with an inside and an outside, but in living membranes it's important to be leaky – but not too leaky.' Such membranes must allow the outward passage of waste products and the inward passage of nutrients. If such microscopic structures do exist in deep space, then they could have offered protection to self-replicating molecules during their long interstellar journeys.

SOME DAMNED DATA

Having examined some of the scientific arguments in favour of the panspermia hypothesis, it is impossible not to give in to temptation and look briefly at a certain strange event which was reported by an ordinary family, and which would seem to suggest that there are some extremely bizarre and incomprehensible things out in space, and that sometimes they fall to Earth. Few professional scientists would accept such reports at face value, and yet they are well worth examining for the hints they present of a bizarre and vividly alive Universe in the midst of which we all live.

The great American anomalist Charles Fort (1874–1932) was the first to bring falls of strange objects from the skies to public attention, and his legacy has been continued by the British journal *Fortean Times*, which began publication in 1973. Fort spent much of his life scouring libraries and periodicals for reports of strange and inexplicable events. He had an evocative phrase for them: he called them 'damned data', by which he meant that orthodox scientists and historians refused to consider them, since they failed to fit comfortably into any existing theory or worldview. Damned by science, these reports were ignored and left in a kind of limbo between existence and oblivion, too inconvenient – too *unsettling* – to be taken seriously, or for time and energy to be spent on a search for their solution. For what could it possibly mean when fish and other animals fell to the earth from out of a clear blue sky? Or bloody lumps of fresh meat?

Or the pulsating, honeycomb-like object that fell into a man's back yard in Miami from out of a clear blue sky in February 1958?

The report of a 'pulsating honeycomb from space' was submitted to the American journal *Fate* by the man who discovered it, Faustin Gallegos, who was then a detective in the Miami Police Department. At about 9.15 a.m. on 28 February, Gallegos was in his living room, discussing the day's shopping itinerary with his wife and mother, when he noticed something strange happening in the family's back yard. As he watched through the two large windows of the living room, a white object 'about the size of a large medicine ball drifted from the sky and made a landing'.

Gallegos asked his wife, Dorothy, who was sitting on the couch facing the windows, if she had seen anything. She replied that something white – a piece of paper, perhaps – had been blown into the yard. His mother, Thelma, had been facing the other way, but said that she'd seen something white reflected in the lenses of her spectacles.

Dorothy went out into the backyard to see what the object was, and immediately called to Thelma and Gallegos to come and see the weird thing that had just arrived. 'I ran out and there at my feet was the strangest-looking substance that I have ever seen,' wrote Gallegos. What he had originally perceived as spherical had now changed its shape so that it looked more like an American football. It was about 20 inches long and 8 inches wide. It seemed to be composed of 'thousands of minute cells resembling those of a honeycomb'. In addition, its colour had changed: no longer was it white, but it had become translucent, like glass . . . and it was pulsating.

After ten minutes spent discussing this strange arrival, Gallegos decided to touch the object – against the strong objections of his mother, who was worried that the thing might be radioactive. But Gallegos' curiosity was too intense, and so he bent down and gently poked the object with his finger. To his astonishment, he felt absolutely nothing: it was as if the object was made of nothing more substantial than air. However, he saw that he had made a hole in the material the size and shape of his finger. 'This was the first time in my life that I had been able to see and touch an object yet been unable to feel it.'

Gallegos then got down on his hands and knees and tried to smell the substance; it was completely odourless. He touched it again, this time drawing his finger along its length, and leaving a deep 'gouge' in its surface.

Dorothy went next door to their neighbour, Mrs Peggy Townsend, and asked her to come and see what had landed in the Gallegos' backyard. Mrs Townsend examined the object, and was as nonplussed as the other three witnesses. While they were discussing it, they saw that the object had begun slowly to shrink, 'as if it were collapsing in upon itself'.

Fearful that the thing would disappear altogether, Gallegos asked his wife to get a jar, so that they could preserve at least part of it. She rushed into the house, and came back with a pickle jar, into which Gallegos scooped part of the material.

> The rest of the object remained upon the ground for a short while, but we saw that it was 'melting' away very rapidly. Soon there was nothing left to show where it had been – not a single trace. We looked over the ground and grass carefully for signs of moisture, imprint, burns, or anything that would show it ever had existed. We could find nothing.

Gallegos turned his attention to the material he had collected in the jar, and noticed that it had stopped pulsating, although it retained its 'cellular' structure. The only explanation he could think of was that this was some kind of 'weather phenomenon'. He called the United States Weather Bureau and described what had happened. The response was not particularly helpful: they had never seen or heard of anything remotely resembling what Gallegos had discovered.

In desperation, the family decided to take the object to his police station; however, the substance began to shrink, and by the time they had made the 20-minute drive to the station, the jar was completely empty.

The *Miami Herald* later ran a story about these curious events, as a result of which the Gallegos residence had many visitors. The place where the object landed was checked for radioactivity, but there was none. As a result of the newspaper report, Gallegos was invited to appear on a radio programme on the subject of UFOs, on which he learned that many other people had seen a similar substance in the past. According to Gallegos:

> A captain of detectives of the Miami Police Department told me, in the presence of an inspector of police, that on the same day the material landed in my yard, an unknown substance also landed in his. Our homes are several miles apart. I believe firmly that both substances are related in body and in origin. I have no idea what the material might be, but I do know it comes from the sky.

Assuming that Faustin Gallegos was telling the truth, what are we to make of the strange interloper that appeared in his back yard on that February morning? Was it some ill-understood terrestrial weather phenomenon, as he initially wondered? It sounds unlikely, given the thing's bizarre properties. Was it really something living, as the pulsations observed suggest? Did it really come from the sky, or rather, *beyond* the sky, in the depths of space of which we know so little? Perhaps these questions will never be answered . . . until another one falls to Earth.

3

ANCIENT CATACLYSMS

THE DEATH OF THE DINOSAURS

Approximately 65 million years ago, something happened to bring to an end the 140-million-year reign of the dinosaurs, and thus to open the way for the rise of mammals and, ultimately, us. However, it wasn't just the dinosaurs that were driven to extinction by this event: about 70 per cent of all the species then living on the planet were also destroyed. It is now thought that, at the time, the dinosaurs were already embarked upon a steady decline towards extinction, and that their demise was merely hastened by the mysterious cataclysm. Others suggest that if the extinction event had not occurred, they might have continued, perhaps evolving into intelligent beings. There is a possibility that if the dinosaurs had not become extinct, the Earth of today might look very different from the world we know!

Many theories have been put forward over the years to account for the extinction of the dinosaurs. These include:

- A giant meteor impact
- The supernova explosion of a nearby star, which bombarded the Earth with lethal radiation
- The birth of too many offspring of a certain gender, which resulted in a catastrophic decline in population
- A disastrous altering of the Earth's climate
- A series of viral infections

Some of these theories are more plausible than others, of course, and their very variety is elegant testimony to the fact that we still don't know for sure just what that terrible extinction event actually was. Whatever happened all those thousands of millennia ago, it resulted in a period of several hundred thousand years during which the dinosaurs and countless other species gradually vanished from the face of the Earth.

THE K-T BOUNDARY

In the late 1970s, the father-and-son team Luis and Walter Alvarez were in Gubbio, Italy, with a group of scientists from the University of California, studying the rocks around the so-called K-T boundary. The letters 'K-T' refer to the transition from the Cretaceous to the Tertiary periods (the 'K' stands for 'Kreta', the Greek word for chalk, which is abundant in rocks from the Cretaceous period). They discovered something very unusual in a layer of clay at the 65 million-year-old boundary point between the Cretaceous and Tertiary: a large concentration of the rare element iridium. In some places, the iridium was 30 times greater than normal levels.

Since there are only two known sources of iridium (the cosmic dust which is constantly falling upon the planet from outer space, and the eruptions of certain types of volcano which expel the element from deep within the Earth), Luis and Walter Alvarez concluded that the anomalously high presence of iridium in the clay could be accounted for only by a massive volcanic eruption or a meteorite strike.

Either of these events would have been truly stupendous in scale: such was the concentration of iridium in the samples studied by the team that they calculated the mass of the postulated meteor at somewhere in the region of 500 million tons. They further calculated that an asteroid of this mass would have been about 10 km in diameter, and would have hit the Earth at a speed of 100,000 kph.

The asteroid impact theory is further supported by the discovery of soot within the clay, which could have been caused by colossal global fires resulting from the high temperatures generated by the impact. In addition to soot, the Alvarezes discovered quartz crystals within the clay; the quartz had been altered in a way that suggested the presence of extreme temperatures and pressures (this type of quartz is known as 'shocked quartz').

This was a plausible and intriguing theory, but the men's peers in the scientific community were still dubious. Such a gargantuan impact should have left an appropriately massive crater, they said. So . . . where was the crater?

THE CHICXULUB IMPACT SITE

In 1990 the scientist Alan Hildebrand was examining some data that had been recorded by geophysicists looking for potential sites for oil exploration in Yucatan, Mexico. As he looked at the data, he was suddenly struck by a curious ring-shaped feature about 180 km in diameter, which lay off the north-west edge of the Yucatan peninsula near the village of Chicxulub.

The creation of this structure was dated at approximately 65 million years ago, and its size and shape were deemed to be consistent with the impact of a 10-kilometre-wide object. Today, the structure cannot be seen without the aid of sophisticated sensing equipment used to detect variations in gravity and magnetic fields, since it is buried beneath several hundred metres of ocean sediment.

But it is there.

The question is: is this the 'smoking gun' that killed the dinosaurs (along with 70 per cent of all life on the prehistoric Earth)?

Such an impact would certainly have done the job. It would have released energy equivalent to millions of tons of TNT, and would have annihilated everything within a radius of 500 km – including the meteorite itself. Massive fires would have been ignited, which would have enveloped the world in a shroud of fire and smoke. Hundreds of billions of tons of debris, including rocks, dust and water vapour, would have been hurled into the upper atmosphere, blotting out the warmth of the sun for years and causing further devastation to plant and animal life. Tidal waves hundreds of metres high would have smashed into coastlines around the world, and would have travelled far inland, causing deluges of truly biblical proportions. In addition, the impact would have triggered massive earthquakes and volcanic eruptions which would have further decimated the land.

Following the lowering of temperatures caused by the 'meteorite winter', there would have been a damaging *increase* in global temperatures, caused by the huge amounts of carbon dioxide released by the worldwide fires. This would in turn have resulted in global acid rains.

The results of this catastrophe would have been dire indeed for most of Earth's life; but the dinosaurs in particular, being cold-blooded, would have suffered greatly from the lowering of temperatures caused by the planet-wide cloud of dust and debris. Food supplies would likewise have been drastically reduced (analysis of the soot in the K-T boundary suggests that as much as a quarter of all plant life was destroyed), resulting in the death of the large herbivores. Of course, with the herbivore population massively reduced, the next to suffer would have been the carnivores as their own food source died out. The only survivors would have been the smaller, more humble animals: tiny scavengers such as the early mammals.

The seas would have fared no better than the land: oxygen levels would have plummeted as lower sections of the oceans, which contain less oxygen, would have been churned up to the surface. The disruption of the marine food chain would have been catastrophic, with much of the plankton dying out, followed by the larger animals that fed on them. The acid rains would have fallen largely into the seas, increasing their acidity and causing further havoc among diverse species.

OBJECTIONS TO THE KILLER ASTEROID THEORY

Although the Alvarezes' theory of a gigantic asteroid strike wiping out the dinosaurs and most other species sounds convincing, some scientists are still unsure as to whether that's what really happened. For one thing, they suggest, the impact of a 10-kilometre-wide asteroid would have caused much *more* damage to life on Earth than appears to have been the case. It sounds strange (and more than a little ironic) to suggest that the damage wasn't serious enough to have been caused by such an object, but some scientists maintain that a 10-kilometre asteroid would have destroyed much, much more than 70 per cent of the Earth's species; in fact, little, if anything, would have survived, and that which did survive would not have had time to evolve and diversify into the several million species we see around us today.

Another problem is raised by the nature of the meteorite winter that would have followed the impact. It would have laid waste to plant and animal species very quickly, in mere centuries or perhaps as quickly as a few decades; and yet we know that the decline of the dinosaurs took approximately 140,000 years. According to some researchers, it took even longer than that.

Two scientists in particular, Professor David Penny of the Allan Wilson Centre for Molecular Ecology and Evolution in New Zealand, and Matt Phillips of the University of Oxford, argue that fossil and molecular evidence do not support the idea that the dinosaurs became extinct suddenly.

While not disputing that a massive asteroid impact did occur 65 million years ago, Penny and Phillips do not accept that it was this event alone which caused the dinosaurs' extinction. While the fossil record is far from complete, and thus prevents unequivocal conclusions from being reached, the revolution in molecular biology can fill in many of the gaps, and scientists can (in theory, at least) reconstruct the family trees of all living things by examining their genes. The evidence presented by Penny and Phillips suggests that the dinosaurs' decline began long before the K-T impact, and continued long afterwards, although the impact may well have contributed to their demise.

NEW EVIDENCE FOR A NEW THEORY

Some scientific theories don't die, they just evolve into new theories. This has happened with the controversy over what, precisely, killed the dinosaurs. According to a recent refinement of the killer impact theory, comets or meteorites hitting the Earth may actually have been responsible for the *beginning* of the dinosaur era.

A team of scientists from several countries has gathered impressive evidence that a massive impact on the prehistoric supercontinent Pangaea 200 million years ago put an end to the Triassic era and brought about the Jurassic. As usually happens in such calamities, more than half of the species on Earth were wiped out; but once everything had settled down and the dust and rock had cleared, a new order of creatures was set to dominate the planet: the dinosaurs.

In an interview with *National Geographic*, Paul Olsen, a geologist at the Lamont-Doherty Earth Observatory of

Columbia University, said: 'We have been able to show for the first time that the transition between Triassic life forms to Jurassic life forms occurred in a geological blink of an eye.' Just as with the K-T impact 140 million years later, the clue to this discovery was a spike in the levels of iridium at the Triassic–Jurassic boundary.

According to Olsen's colleague, Dennis V. Kent of Rutgers University, who co-authored a paper on the subject with Olsen, several different strands of evidence were gathered, including 'dinosaur footprints, skeletal remains, and a spike in iridium levels exactly at the change in plant spores from the strata that mark the Triassic–Jurassic boundary', all suggesting that the Earth was hit by a giant comet or asteroid 200 million years ago.

It should be remembered that, at least on geological timescales, mass extinction events happen fairly regularly; in fact, our planet has suffered no fewer than five such events during its four-and-a-half-billion-year history. In the aftermath of each extinction event, new species appear and thrive. The Triassic–Jurassic event led to the rise of the dinosaurs; and the Cretaceous–Tertiary event led to the rise of mammals and birds.

The Triassic–Jurassic boundary is marked by a 'fern spike', according to Kent. Ferns are very hardy plants which are always the first to colonise an area that has been devastated by some disaster (this can be seen from the abundance of ferns on the slopes of Mount St Helens in Washington, USA, following its eruption). Any increase in fern spores strongly implies an 'ecological calamity'.

Just such an increase was observed in rock strata from the so-called 'Newark Basin', which was formed in distant epochs just prior to the separation of North America from Europe and Africa. The geologists also discovered an iridium spike in these strata, which, as we have seen, implies a massive depositing of the element from deep space in the form of an asteroid impact. This iridium spike was discovered thanks to new techniques of mass spectrometry, which allow measurement in the parts per trillion.

Olsen and Kent discovered something else just below the Triassic–Jurassic boundary: a thin layer of sediment where the Earth's magnetic field is reversed (a phenomenon that happens at regular intervals, and which can be used to date events in the distant past).

The final piece of evidence was discovered just above the Triassic–Jurassic boundary. At that time, there were massive lava flows that occurred across huge tracts of land. These flows, the geologists suggest, were triggered by the impact from space.

According to Hans-Dieter Sues, a paleobiologist at the Royal Ontario Museum in Canada:

> There are similarities to the K-T extinction (Cretaceous–Tertiary Boundary, 65 million years ago). The fern spike indicating a significant ecosystem disruption, the disappearance of large groups of large land animals, higher concentrations of iridium. All these different pieces of evidence build a picture supporting the idea that a collision occurred.

The early dinosaurs came through the catastrophe relatively unscathed, but their major competitors didn't; and so the dinosaurs became the planet's dominant life form. With so many of the earlier animals eliminated, the dinosaurs no longer had to compete for food. This also accounts for their rapid spread across the globe, and probably their increase in size.

For the next 140 million years, the dinosaurs ruled the Earth . . . until they also began to decline, perhaps due to the next massive impact at the Cretaceous–Tertiary boundary.

THE SIGNIFICANCE OF CLIMATE

Some scientists are less concerned with massive impacts from deep space than with the natural cycles of the Earth's climate over long periods. They suggest that it may be more sensible to look in this direction, both in terms of an answer to what killed the dinosaurs, and in terms of our own future on the planet.

There is, for instance, a theory that an extinction cycle exists which repeats every 60 to 70 million years. The cause of this cycle is still uncertain: it could be due to volcanic action, or to a periodic drop in sea level. Scientists at NASA and the United States Geological Survey have been using the US GOES satellite system to compare weather and water levels over periods of millions of years.

Their findings suggest that there was a highly significant drop in sea levels about 65 million years ago. Such a drop would undoubtedly have resulted in significant climatic change across the world. Any species unable to adapt to this change quickly enough would have been doomed. Higher temperatures and increased carbon dioxide might possibly have spelled the end for the dinosaurs, allowing mammals and other more adaptable animals to evolve and thrive in the new environment.

It is, of course, rather unsettling to contemplate the suggestion that these extinction events occur every 60 to 70 million years. After all, the last one was 65 million years ago. By a curious coincidence, astronomers believe that asteroid impacts of the magnitude of the K-T impact also occur every 60 to 70 million years, and the last one of *those* was 65 million years ago.

Sometimes, one wonders whether it's really worth getting out of bed in the morning …

4

STRANGE TRACES

ANOMALOUS FOSSILS

In broad terms, the timeline of life on Earth would appear to be fairly straightforward. Although it contains many gaps, mysteries and archaeological conundrums, the story we have been able to piece together from palaeontological and archaeological evidence seems to describe an aeon-long movement from simplicity towards complexity, beginning with the first self-replicating molecules of the 'primeval soup' on the young Earth, to humanity's present technological civilisation.

The same goes for the history of our species, *Homo sapiens*. We can trace its development from the origin of bipedalism in Africa approximately five million years ago, through the era of brain expansion and the first use of tools, to the period about two million years ago when *Homo erectus* began to expand out of Africa and into Asia. The period between one and two million years ago saw the cultural period known as the Achulean, during which there were major advances in tool manufacture, and meat-eating became more widespread. The first use of fire seems to have occurred about 700,000 years ago; further major advances in tool manufacture occurred during the Mousterian period 200,000 years ago; modern humans appeared in Africa at about the same time; the first evidence of representational art appeared 30,000 years ago; the Agricultural Revolution took place 10,000 years ago and the first cities appeared 5,000 years ago; the Industrial Revolution, which made the modern world possible, occurred 150 years ago.

It all looks pretty straightforward . . . until one begins to consider certain other archaeological discoveries made over the last 100 years or so. These discoveries, which have been either derided or ignored by orthodox archaeologists, tell a different story of the remote history of our world. It is a story that is astonishing in its implications, for it suggests that there may have been humans (or something very like humans) walking the Earth as long as 300 million years ago.

'LOOKS HUMAN: REMARKABLE'

One cannot really blame archaeologists for throwing up their hands in horror at the very idea of creatures similar to modern humans walking around millions of years *before* the dinosaurs. Many children have watched in wonder such entertaining fantasies as the Hammer classic *One Million BC*, in which Raquel Welch (clad in a fetching animal-skin bikini) battles carnivorous dinosaurs, only to be told by their smiling parents that the great lizards became extinct millions of years (about 65 million, to be more or less exact) before the first humans walked the Earth. The idea that humans *pre-dated* the dinosaurs is even more ridiculous, since it flies in the face of all we have learned about the early history of our world, and the development of our species.

How, then, to explain the discoveries made by Dr Wilbur Greely Burroughs, head of the Geology Department of Berea College in Kentucky, during his expedition to the Kentucky hills in the 1930s? Burroughs came upon no fewer than ten man-like footprints in carboniferous sandstone, which had apparently been made by something walking upright across a sandy beach in the Pennsylvanian Period of the Palaeozoic Era. The tracks looked like they had been made by a man.

A man who lived approximately 250 million years ago.

Being a properly cautious scientist, Dr Burroughs at first could not believe that he had found evidence of a living human being walking around a quarter of a billion years before he should have existed. For seven years he kept his discovery secret, and devoted himself to further study of the astonishing footprints.

Even when he finally decided to go public, he maintained a non-committal stance, refusing to state outright that the tracks were made by what seemed to be a modern human, and preferring to confine himself to a straightforward description of what he had found. In an interview with the Louisville *Courier-Journal* magazine some 20 years after the discovery, Burroughs stated:

> Three pairs of tracks show both left and right footprints. Of these, two pairs show the left foot advanced relative to the right. The position of the feet is the same as that of a person. The distance from heel to heel is 18 inches. One pair shows the feet about parallel to each other, the distance between the feet being the same as that of a normal human being.

Burroughs's findings seemed irrefutable: the footprints displayed a heel and five toes, and strongly implied a bipedal creature. According to the famous anomalies researcher Brad Steiger, in his book *Mysteries of Time and Space*, Dr Burroughs followed the suggestion of his colleague, Dr Frank Thone, and, with the agreement of Charles Gilmore of the Smithsonian Institute, he named the creature that had left the tracks *Phenanthropus mirabilis* ('looks human; remarkable').

Objections were, of course, raised by other scientists in the field, who suggested that the tracks might have been made by one of the large amphibians that existed during the Pennsylvanian Period. However, Dr Burroughs countered thus: 'There is no indication of front feet, though the rock is large enough to have shown front feet if they had been used in walking.'

Others suggested that the marks might have been made by Amerindian artists, who could have carved them into the rock. Dr Burroughs sought the opinions of sculptors, who told him that such carvings would include the tell-tale markings of the sculpting instruments – none of which were evident.

According to Steiger:

> The cautious scientist meticulously examined what appeared to be a ridgelike roll around the edge of the more deeply imbedded footprints, indicating that pressure beneath the creature's foot had moved the loose sand a bit higher than the surrounding surface. With the aid of a compound microscope, Dr

Burroughs counted the grains of sand for a unit of area, just as a medical doctor would count the red or white corpuscles in a specimen of blood.

Dr Burroughs found that the sand grains inside each track were packed closer together than those surrounding the tracks. This, he concluded was the result of the grains being pressed together by the weight of the creature.

If barefoot humans walking around 250 million years ago sounds bizarre and unbelievable, the idea that other humans were living even further back in time – 600 million years or so – and furthermore were wearing shoes, is surely beyond the bounds of all reason. And yet that would appear to be precisely what was discovered in the state of Utah in 1968.

A CRUSHED TRILOBITE

The man who made the discovery, William Meister, was a keen amateur fossil-hunter – a 'rockhound', to use his own expression. On 1 June 1968, Meister was at Antelope Springs, 43 miles northwest of Delta, Utah, with his wife and two daughters. Also present were their friends, Mr and Mrs Francis Shape, and their two daughters.

On the third day of their four-day sojourn at Antelope Springs, Meister discovered some trilobite fossils. He was particularly fascinated by these ancient, long-extinct arthropods, and so was very glad to make this find. He continued his search, hoping to uncover more trilobites, which are among the oldest-known fossils. Breaking off a large, two-inch-thick slab of rock, Meister gave it a tap with his hammer, splitting it along its length so that he was left with two one-inch-thick slabs.

Inside was something which at first he could not credit, could not even understand until he examined it more closely and realisation dawned upon him. Meister described it in an article that appeared in the journal *Creation Research Quarterly*, in the December 1968 issue. '[I saw] on one side the footprint of a human with trilobites *right in the footprint itself*. The other half of the rock slab showed an almost perfect mold of the footprint and fossils [original emphasis].' Meister realised that, incredibly, the human had been wearing a sandal.

The footprint was 10$^1/_4$ inches long and 3$^1/_2$ inches wide at the sole, and evidently had made a deeper depression on its right side, implying that it was a right foot. Within the footprint were the clearly distinguishable forms of several small trilobites, which led Meister to the astonishing conclusion that whoever had worn the shoe had trodden upon the creatures while walking, and had done so during the Cambrian Period of the Palaeozoic Era . . . 600 million years ago. This individual had done nothing so prosaic as 'walking with dinosaurs': he or she had walked the land 300 million years before the first dinosaurs appeared!

Assuming that Meister's discovery was indeed what it appeared to be, who was the individual who had lived so very long before human beings are supposed to have lived? Was the individual human? Or did he or she belong to an unknown and long-vanished human-like race whose fossils and other artefacts have yet to be discovered?

A 500,000-YEAR-OLD SPARKPLUG?

Seven years before William Meister made his discovery of an apparently shod human foot in a 600-million-year-old fossil, three people set out for the Coso Mountains north-east of Olancha, California. Mike Mikeshell, Wallace Lane and Virginia Maxey were hoping to find some semi-precious minerals for their LM & V Rockhounds Gem and Gift Shop. As Brad Steiger portentously notes: 'Instead, they found what may be a clue to a precataclysmic world.'

While examining some stones near the summit of a peak above the dry bed of Owens Lake, they came upon something that, at first sight, looked like an ordinary geode, its surface encrusted with fossil shells. The following day, Mike Mikeshell was in the gift shop, cutting the geode in half with his 10-inch diamond-tipped saw, when the blade struck something inside the stone. Opening it, Mikeshell discovered that it did not contain an empty cavity, as he had been expecting; instead, he found something altogether more peculiar: a circular section of a very hard, ceramic-like material containing a two-millimetre-diameter shaft of metal running through its centre.

Steiger reports that the rock-hunters also discovered two non-magnetic metallic objects, resembling a nail and a washer, inside the geode. There was also a hexagonal casing around the ceramic disc. 'The metal core of the disc responded to a magnet. In the opinion of Mike Mikeshell, Virginia Maxey and Wally Lane, there appeared to be some evidence that the ceramic core had been encased in copper, for a bit of the metal seemed intact while the rest had decomposed.'

What was this strange, apparently technological artefact? Sceptics have suggested that it was merely an artefact from our own modern world, something that had fallen into some mud that had then been baked hard in the sun. However, this does not address the fact that the artefact was found inside a stone covered with fossil shells, which, according to a geologist whom the trio consulted, would have taken at least 500,000 years 'to attain its present form'.

The artefact (and the X-ray photographs that were taken of it) came to the attention of Paul Willis, editor of the UFO magazine *INFO Journal* in 1969. After pondering the images for some time, Willis suddenly realised that the object resembled nothing so much as a sparkplug.

Willis told his brother, Ron, who wrote an article on the subject for the *INFO Journal*:

> I was thunderstruck, for suddenly all the parts seemed to fit. The object sliced in two shows a hexagonal part, a porcelain or ceramic insulator with a central ceramic shaft – the basic components of any sparkplug.

The Willis brothers then took apart an ordinary sparkplug and examined its components.

> We found all the components similar to the Coso artifact, but with some differences. The copper ring around the halves displayed in the object seems to correspond to a copper sealer ring in the upper part of the steel casing of any sparkplug.

The Willis brothers then contacted Wallace Lane and asked to examine the artefact itself. However, he replied that it could be purchased for $25,000, and several museums were already interested.

Ron Willis concludes his article in the *INFO Journal* with the rather rueful observation that:

> There is no indication that any professional scientist has ever carefully examined the object, so what it may be is still questionable. The Coso artifact now seems to join the club with the Casper, Wyoming mummy . . . and other Fortean objects whose owners refuse to allow anyone to examine the object in question without an exorbitant payment.

'PEDRO'

What was the 'Casper, Wyoming mummy' to which Ron Willis referred in his article? Simply put, it is one of the most curious objects ever found, and may hold clues to the existence of unknown human or humanoid groups who once shared (and perhaps still share) the planet with us.

In October 1932, in the Pedro Mountains about 100 km west of the town of Casper, Wyoming, two gold prospectors were exploring a gulch at the base of the mountains, when they came across what they thought was an indication of gold in one of the walls. After setting explosive charges in the wall and blasting away part of it, they saw that they had exposed a cave about a metre square and five metres deep. Inside the cave, they discovered something that should not have been there: the tiny, desiccated figure of a man sitting cross-legged on a ledge. His arms were also crossed, his body was dark bronze in colour and he was only 35 cm high.

The two astounded prospectors promptly took their curious find back to Casper. The initial mutterings of 'hoax' were dispelled when the tiny, mummified man was X-rayed by anthropologists, and was found to contain a skull, spine, ribcage and other bones, virtually identical to those found in a normal human. The mummy also had a full set of teeth, weighed about 340 grams and, by the anthropologists' estimates, was about 65 years old when he died. According to the Wyoming State Historical Society, the tiny man was undoubtedly human – a view shared by anthropologists from Harvard.

Dr Henry Shapiro of the American Museum of Natural History declared that the mummy was 'of an extremely great age, historically speaking, and of a type and stature quite unknown to us'. The mummy, which had been nicknamed 'Pedro' after the mountains where he had been discovered, was also examined by the Boston Museum Egyptian Department, which supported the earlier contention that it was a fully grown adult, not a child, and also that the method of preservation seemed to match that of the Egyptian pharaohs.

Could it be that the two gold prospectors had stumbled on the burial place of an ancient and (presumably) long-dead humanoid race? Certainly, although they declared the Casper mummy to be genuine, the anthropologists who investigated the case were somewhat reluctant to commit themselves as to its possible origin.

THE MYSTERIOUS INHABITANTS OF FLORES: PEDRO'S COUSINS?

Pedro, the Wyoming mummy, has always been considered one of those bizarre little historical and scientific puzzles that will almost certainly never be solved, an odd aberration that, taken in isolation, can add very little to an understanding of humanity's history.

At least, that was the case until 2004, when certain discoveries by Australian and Indonesian archaeologists shed new light on the mystery of Pedro – although few if any commentators have made the connection. The scientists have discovered remains on the Indonesian island of Flores, which suggest that a completely different and hitherto unknown species of human being (with physical features normally associated with hominids from 1.5 and 4 million years ago) was living in the region until a mere 12,000 years ago.

The skeletal remains of seven individuals were found by scientists from Australia's University of New England and University of Wollongong, in a cave at Liang Bua, Flores. The three-foot-tall skeletons have the small brain cases (380 cc) and receding chins associated with *Australopithecus*, a hominid which was thought to have lived more than three million years ago. However, other features of the skulls, such as their relatively flat faces and small molar teeth, are more in keeping with *Homo erectus*, a late ancestor of modern humanity.

Although the creatures have been given the sober binomial classification *Homo floresiensis*, media journalists were quick to describe them as 'the real hobbits', after the diminutive, nature-loving people in J.R.R. Tolkien's classic fantasy *The Lord of the Rings*. They might just as well have called them 'Pedro's cousins', although admittedly that doesn't quite have the same ring to it, and would have resulted in a great deal of head-scratching among those who have never heard of the Wyoming mummy!

According to the journalist David Keys, who reported on the find for the Archaeological Institute of America, it is possible that *Homo floresiensis*' small stature could be the result of living on a small island for hundreds of millennia, where the absence of predators made large size redundant as a defensive advantage.

The puzzling thing about the discoveries on Flores is that they include sophisticated stone tools, including cutting blades and spear points, which are comparable to those made by Stone Age humans, and yet *Homo floresiensis*' brain capacity bears a greater resemblance to hominid groups, such as *Australopithecus*, who fashioned only very rudimentary tools. The obvious answer would appear to be that the tools found in the cave at Liang Bua were indeed made by Stone Age humans, and not by *Homo floresiensis*. However, the earliest of the tools found in the cave date from 90,000 years ago, and *Homo sapiens* only arrived in the region about 50,000 years ago. The implication is that this strange, previously unknown ancestral group really did make these sophisticated tools.

In an intriguing aside, David Keys notes that there are certain folktales in the region which hint at the astounding possibility that *Homo floresiensis* may have survived into the nineteenth century. These tales, which are still told by the villagers on Flores, describe the existence of small, fur-covered people who used to steal food from them many years ago. They were known as the *ebu gogos* (meaning 'the grandmothers who eat anything'). Bert Roberts, a member of the excavation team from the University of Wollongong, expressed his belief that the folktales of the present inhabitants of Flores raise the possibility not only that *Homo floresiensis* may have survived into the nineteenth century, but also that 'they still survive today in some remote jungle area of the island'.

When we compare these tales to those of the *orang pendek*, a four-foot-tall, hairy ape-like creature said to haunt the forests of nearby Sumatra, we may ask ourselves whether there might be yet more survivals from a remote history whose wonders we have barely begun to explore.

5

THE REMAINS OF LEMURIA?

THE ENIGMA OF THE YONAGUNI MONUMENT

Most people are aware of the legend of Atlantis, the ancient civilisation that disappeared beneath the ocean's waves in a single cataclysmic day and night. Rather less familiar is the legend of Lemuria, an equally vast and mysterious island-continent that is believed to have been located somewhere in the Pacific (or perhaps Indian) Ocean, and which, like Atlantis, sank beneath the ocean's surface in the distant past, leaving only a few scattered islands as testament to its existence.

However, unlike Atlantis, which is first mentioned by Plato in the fourth century BC, the concept of Lemuria is a modern invention – although, as we shall see, it has been appropriated by occultists and pseudo-scientific writers who believe that it was once home to an unknown and highly advanced prehistoric civilisation. The name 'Lemuria' was coined in 1864 by the zoologist Philip Sclater in an article entitled 'The Mammals of Madagascar', which was published in *The Quarterly Journal of Science*.

Sclater had been intrigued by the presence of fossil lemurs in Madagascar and India and surrounding regions, but nowhere else in the world. Why, he wondered, were the remains of those beautiful and enigmatic primates found only in isolation, when they should have been much more widely distributed? He decided that the answer had to be that India and Madagascar had once been part of a much larger land mass; and since his line of reasoning had been inspired by the mystery of the lemurs, he called the long-vanished continent 'Lemuria'.

Sclater's idea was eminently sensible within the context of scientific knowledge at the time. In the mid-nineteenth century, with the rise and acceptance of Darwinism, scientists attempted to trace the development of various species from what were believed to be their points of origin; and since these points were often separated by bodies of water, it seemed natural to assume that they had once been connected by land bridges which had later become submerged.

The concept of Lemuria gained widespread acceptance within the scientific community, with some (notably the German Darwinian Ernst Haeckel) seizing upon it as a possible explanation for the absence of the so-called 'missing link' between apes and modern humans.

However plausible the theory of sunken prehistoric land masses might have been at the time, it could not survive the discovery of plate tectonics and continental drift, which explained the distribution of lemurs and many other species. In addition, it discredited the idea that large tracts of land could quickly become submerged beneath the ocean, since this is simply not the way the continental plates behave.

A LEGEND TO RIVAL ATLANTIS

Having been abandoned by scientists, the concept of the ancient, lost world of Lemuria was enthusiastically picked up by occultists in the late nineteenth century, most notably Madame Helena Petrovna Blavatsky (1831–91), who founded the Theosophical Society in 1875.

In 1877, Blavatsky published a book entitled *Isis Unveiled*, an exposition of Egyptian occultism that, she claimed, had been dictated to her by spirits via a form of automatic writing, and which argues for the acceptance of occultism (i.e. the study of the hidden laws of nature) by orthodox science. The book sold widely, and had the effect of soothing the minds of those whose religious faith had been undermined by scientific rationalism, in particular the theory of evolution and natural selection developed by Charles Darwin. The book was fiercely attacked in scholarly circles for both intellectual incompetence and plagiarism, with one critic identifying more than 2,000 unacknowledged quotations.

Central to the personal myths Blavatsky constructed for herself was her experience of living and travelling for seven years in Tibet. She made the astonishing claim that she had studied with a group of 'Hidden Masters' in the Himalayas, under whose guidance she had reached the highest level of initiation into the mysteries of the Universe. However, as Peter Washington notes in his book *Madame Blavatsky's Baboon: Theosophy and the Emergence of the Western Guru*, it is extremely unlikely that a single white woman with a considerable weight problem and no mountaineering experience could have made the arduous trip to the Himalayas, succeeded in finding these Hidden Masters, and done so without being spotted and apprehended by the numerous Chinese, Russian and British patrols that were in the area at the time.

In 1888, Blavatsky published *The Secret Doctrine*, a vast work comprising two main sections, 'Cosmogenesis' and 'Anthropogenesis', which claims to be nothing less than a history of the Universe and of intelligent life. The book is allegedly a commentary on a fantastically old manuscript called *The Stanzas of Dzyan*, written in the Atlantean language 'Senzar', which Blavatsky claimed to have seen in a monastery hidden far beneath the Himalayas. The *Stanzas* tell how Earth was originally colonised by spiritual beings from the Moon. Modern humanity is descended from these remote ancestors via a series of so-called 'root races'.

According to Blavatsky's cosmology, at the beginning of the Universe the divine being differentiated itself into the multitude of life forms that how inhabit the cosmos. The subsequent history of the Universe passed through seven 'rounds' or cycles of being. The Universe experienced a fall from divine grace through the first four rounds, and will rise again through the last three, until it is redeemed in ultimate divine unity, before the process begins again.

On Earth, each of these cosmic rounds saw the rise and fall of seven root races, whose destiny exactly mirrored that of cosmic evolution, with the first four descending from the spiritual into the material, and the last three ascending once again. According to Blavatsky, humanity in its present form is the fifth root race of Earth, which is itself passing through the fourth cosmic round. (The reader may thus be relieved to learn that we have a long period of spiritual improvement ahead of us.)

The first root race was composed of completely non-corporeal astral beings who lived in an invisible land. The second race were the Hyperboreans, who lived on a lost polar continent. The third root race lived on the island-continent of Lemuria, and were 15-foot-tall brown-skinned hermaphrodites with four arms, who had the misfortune to occupy the lowest point in the seven-stage cycle of humanity. For this reason, the Lemurians suffered a fall from divine grace: after dividing into two distinct sexes, they began to breed with beautiful but inferior races, and this miscegenation resulted in the birth of soulless monsters.

The fourth root race were the Atlanteans, who possessed highly advanced psychic powers and mediumistic skills. Gigantic like the Lemurians and physically powerful, the Atlanteans built huge cities on their mid-Atlantic continent. Their technology was also highly advanced, and was based on the application of a universal electro-spiritual force known as 'Fohat'. Unfortunately for the Atlanteans, although they were intelligent and powerful, they were also possessed of a child-like innocence that made them vulnerable to the attentions of an evil entity that corrupted them and caused them to turn to the use of black magic. This resulted in a catastrophic war which led in turn to the destruction of Atlantis. The fifth root race is modern humanity, with the sixth and seventh races still to come at some unspecified time in the future.

In the twentieth century, the concept of Lemuria was further developed by occultists, fringe scientists and science fiction writers. One of the most famous of the many legends that have grown up around Lemuria and the survivors of the catastrophe which overwhelmed the island-continent in the distant past concerns the majestic and beautiful Mount Shasta in northern California.

In recent years, fringe writers have made outlandish and totally unsubstantiated claims regarding the civilisation existing inside Mount Shasta, which has become something of a Mecca for New Agers from all over North America and the rest of the world. In particular, the American writer Timothy Green Beckley, whose books are weird and wonderful flights of cosmic fancy, claims that there is a city called Telos hidden inside the mountain, and which is the 'primary Lemurian outpost . . . with a small secondary city in Mount Lassen, California'.

Telos, writes Beckley, contains five levels and is home to one and a half million Lemurians, some of whom occasionally venture forth, dressed in white robes, to walk in the mountain's foothills. The uppermost level contains residential districts and a spaceport (!), and is devoted to education and administration. Lower levels contain hydroponic gardens for food production, manufacturing centres and subterranean parks. It is apparently the wish of the Lemurians that one day open contact will be made with surface humanity, and that the two peoples will co-exist in peace and mutual understanding.

While all this is highly reminiscent of the pulp science fantasy of the 1930s and '40s (and bears about as much critical examination), the question of whether 'Lemuria' really existed remains. As we shall see, recent discoveries off the coast of Japan have re-invigorated the concept of ancient sunken lands, and are making some archaeologists wonder whether at least part of the ancient history of our species will have to be rewritten.

THE YONAGUNI ENIGMA

Dr Robert M. Schoch of Boston University's College of General Studies has conducted detailed examinations of several of the world's great monuments, including the Great Sphinx at Giza, and has been extremely courageous in making his conclusions public. For instance, his radical re-dating of the Sphinx by approximately 8,000 years, based on the monument's weathering patterns (which Schoch concluded could only have been created by rainfall), has caused consternation among orthodox Egyptologists.

Like many others, Schoch is fascinated by the idea of Atlantis and other 'lost civilisations', and so it was with alacrity that in 1998 he accepted the invitation of a Japanese businessman, Yasuo Watanabe, to examine a recently discovered underwater structure off the coast of Yonaguni Island east of Taiwan and west of Ishigaki and Iriomote Islands in the East China Sea.

In his paper 'An Enigmatic Ancient Underwater Structure Off the Coast of Yonaguni Island, Japan', Schoch states that the

'Yonaguni Monument' has the appearance of 'a platform-like or partial step-pyramid-like structure'. He continues:

> The Yonaguni Monument is over 50 metres long in an east–west direction and over 20 metres wide in a north–south direction. The top of the structure lies about 5 metres below sea level, whereas the base is approximately 25 metres below the surface. It is an asymmetrical structure with what appear to be titanic stone steps exposed on its southern face. These steps range from less than half a metre to several metres in height.

Noting that the rock faces appear to be dressed stone, Schoch suggests that if this is indeed an artificial structure, it would be logical to assume that it was fashioned at a time when the area was above sea level. Since the area experienced sea level rises following the last Ice Age, 'we can suggest with some confidence that if the Yonaguni Monument is a man-made construction then it must be at least 8,000 years old'.

After meeting with Dr Masaaki Kimura of the Department of Physics and Earth Sciences at the University of the Ryukyus, Okinawa, whose examination of the structure led him to conclude that it had indeed been fashioned by human hands, Schoch made several dives during which he also examined the structure. He scraped off the abundant organisms (including algae and corals) that had made their home on the vast stone surfaces, and took rock samples which proved to be of 'very fine sandstones and mudstones of the Lower Miocene Yaeyama Group' which had been deposited some 20 million years ago.

After noting the parallel bedding planes which result in the separation of layers, Schoch decided to examine similar sandstones on Yonaguni Island itself. He observed that very similar 'natural, but highly regular' features occurred as a result of natural erosion of the stones above sea level. On the surface, he also 'found depressions and cavities forming naturally that look exactly like the supposed "post holes" that some researchers have noticed on the underwater Yonaguni Monument'.

In addition, upon closer examination, Schoch could find no evidence that the underwater stones had been worked with tools, or moved into place (although he adds that just because he has not found such evidence, that doesn't mean it doesn't exist). Although he concluded that the Yonaguni structure is the result of natural processes, Schoch leaves the door open on the question of whether it was later modified and used for some unknown purpose by human beings. He notes that there is a tradition in the region of 'modifying, enhancing and improving on nature'. On Yonaguni Island, he notes, there are some ancient tombs of unknown provenance which bear a certain resemblance to the underwater structure, and he further speculates that 'even if it is a primarily natural structure, it may have been reshaped to serve as foundation for stone, timber, or mud buildings that have since been destroyed'.

In view of Schoch's conclusions, Dr Kimura subsequently modified his own opinion on the origin of the Yonaguni structure, suggesting that it had been 'terraformed' – that is, modified by human hands. This stance is more palatable to Schoch, who maintains that, while probably of natural origin, the Yonaguni structure is certainly worthy of continuing investigation.

It seems that Yonaguni is not the unequivocal proof of an unknown and highly advanced human civilisation in prehistoric times that many have been hoping for. We will have to keep looking for Lemuria ...

6

LIFE FROM LIFELESSNESS?

THE ACARI OF ANDREW CROSSE

The people of the area shunned the dilapidated Fyne Court estate, believing it to be the home of ghosts and devils. They were especially mindful to avoid the place after nightfall, when strange lights danced upon the metal wires that had been strung around the grounds, and horrible crackling sounds emanated from the house to disturb the peace and quiet of this normally tranquil region of the Quantock Hills in Somerset. The locals called the reclusive owner of the estate 'the thunder and lightning man', and complained among themselves that whatever he was doing, it could only bode ill for him, and perhaps for them also: for his activities, they believed, must be unnatural, an affront to God and to human decency. Before long, they came to call him by another nickname: the 'Wizard of the Quantocks'.

The year was 1836, and within 12 months the fear and mistrust with which the owner of Fyne Court was regarded had spread throughout the country. He was denounced as a worker of the foulest blasphemies, 'a reviler of our holy religion', a man whose atheistic arrogance and presumption had tempted him to take on the role of God Himself as the creator of life. He became the archetypal 'mad scientist', a real-life Frankenstein, to be hated and shunned by all right-thinking, God-fearing people.

Who was the owner of Fyne Court, and what was the exact nature of his terrible crime?

The man's name was Andrew Crosse, and the 'crime' of which he was accused was the creation of living creatures through the application of electricity to non-living matter.

Born on 17 June 1784 at the family seat in Broomfield, Somerset, Crosse quickly became acquainted with tragic loss. While still a child, he lost his father, sister, an uncle and two of his best friends; and in 1805 his mother's death left him an orphan at the age of 21. He had completed his education, first at Dr Seyer's School in Bristol, and then at Brasenose College, Oxford, and from then on he lived a solitary life at Fyne Court, where he continued with his studies of electricity, chemistry and mineralogy – subjects that had fascinated him since childhood.

At this point in his life he was not a total recluse: he had some friends, including George Singer, whose book *Elements of Electricity and Electro-Chemistry* was published in 1814. Crosse's own experimental work began in 1807, when he embarked upon the study of the formation of crystals through the application of electrical currents (he had, it seems, been inspired by the stalactites and stalagmites in Holywell Cavern at Broomfield).

In 1809 he married Mary Anne Hamilton; the couple had seven children, three of whom died at birth. At that time in his life, Crosse had yet to become completely reclusive, and he delivered several lectures on his work, one of which, given in December 1814 in London, was attended by Percy Bysshe Shelley and his young companion, Mary Wollstonecraft Godwin. Mary, of course, would go on to write one of the greatest and most famous novels of gothic horror, *Frankenstein*; and it is possible – perhaps likely – that she modelled her tragic hero, Victor Frankenstein, on Andrew Crosse and his experiments with atmospheric electricity.

Following the death of his friend, George Singer, in 1817, Crosse became increasingly reclusive, devoting all his time to his scientific research. He strung more than a mile of copper wire on poles through the grounds of Fyne Court, which fed into his 'electrical room', and by means of which he attempted to study the electricity in the atmosphere (it was this network of wire that caused such trepidation in the local inhabitants of Broomfield).

His astonishing breakthrough (if such it was) occurred in 1837, when he was conducting certain experiments on the artificial formation of crystals by means of weak but long-applied electric currents. He was trying to produce crystals of silica by allowing a suitable fluid medium to seep through a piece of porous stone, while applying an electric current from a voltaic battery. The fluid was a mixture of hydrochloric acid and a solution of silicate of potash.

According to Crosse himself:

> On the fourteenth day from the commencement of this experiment I observed through a lens a few small whitish excrescences or nipples, projecting from about the middle of the electrified stone. On the eighteenth day these projections enlarged, and stuck out seven or eight filaments, each of them longer than the hemisphere on which they grew.
>
> On the twenty-sixth day these appearances assumed the form of a *perfect insect*, standing erect on a few bristles which formed its tail. Till this period I had no notion that these appearances were other than an incipient mineral formation. On the twenty-eighth day these little creatures moved their legs. I must now say that I was not a little astonished. After a few days they detached themselves from the stone, and moved about at pleasure.

Over the course of the next few weeks, according to Crosse, about 100 of the creatures appeared on the stone. He examined them closely under a microscope, and saw that the smaller ones appeared to have six legs, and the larger ones eight. He decided that they must be of the genus *acarus* (i.e. mites), but wondered whether they were a known species, or one never before seen.

> I have never ventured an opinion on the cause of their birth, and for a very good reason – I was unable to form one. The simplest solution of the problem which occurred to me was that they arose from ova deposited by insects floating in the atmosphere and hatched by electric action. Still I could not imagine that an ovum

could shoot out filaments, or that these filaments could become bristles, and moreover I could not detect, on the closest examination, the remains of a shell . . .

I next imagined, as others have done, that they might originate from the water, and consequently made a close examination of numbers of vessels filled with the same fluid: in none of these could I perceive a trace of an insect, nor could I see any in any other part of the room.

Crosse then modified his experiments, discarding the porous stone, and found that he could produce the *acari* in glass cylinders filled with concentrated solutions of copper nitrate, copper sulphate and zinc sulphate. The creatures usually appeared at the edge of the fluid surface; however, Crosse added that 'in some cases these insects appear two inches *under* the electrified fluid, but after emerging from it they were destroyed if thrown back'. (It should be noted that if these creatures really were *acari*, then they were not insects but arachnids.)

In one experiment, the *acari* appeared on a small piece of quartz, immersed at a depth of 2 inches in fluoric acid holding silica in solution.

A current of electricity was passed through this fluid for a twelvemonth or more; and at the end of some months three of these *acari* were visible on the piece of quartz, which was kept negatively electrified. I have closely examined the progress of these insects.

Their first appearance consists in a very minute whitish hemisphere, formed upon the surface of the electrified body, sometimes at the positive end, and sometimes at the negative, and occasionally between the two, or in the middle of the electrified current; and sometimes upon all. In a few days this speck enlarges and elongates vertically, and shoots out filaments of a whitish wavy appearance, and easily seen through a lens of very low power.

Then commences the first appearance of animal life. If a fine point be made to approach these filaments,

they immediately shrink up and collapse like
zoophytes upon moss, but expand again some time
after the removal of the point. Some days afterwards
these filaments become legs and bristles, and a perfect
acarus is the result, which finally detaches itself from
its birthplace, and if under a fluid, climbs up the
electrified wire and escapes from the vessel ...

If one of them be afterwards thrown into the
fluid in which he was produced, he is immediately
drowned . . . I have never before heard of *acari*
having been produced under a fluid, or of their ova
throwing out filaments; nor have I ever observed any
ova previous to or during electrization, except that
the speck which throws out filaments be an ovum;
but when a number of these insects, in a perfect state,
congregate, ova are produced.

In a later experiment, Crosse managed to produce an *acarus* in
a closed and airtight glass retort, filled with an electrified silicate
solution. On connecting the battery, Crosse observed that:

An electric action commenced; oxygen and hydrogen
gases were liberated; the volume of atmospheric air
was soon expelled. Every care had been taken to avoid
atmospheric contact and admittance of extraneous
matter, and the retort itself had previously been
washed with hot alcohol. This apparatus was placed
in a dark cellar.

I discovered no sign of incipient animal formation
until on the 140th day, when I plainly distinguished
one acarus actively crawling about *within* the bulb
of the retort.

I found that I had made a great error in this
experiment; and I believe it was in consequence of this
error that I not only lost sight of the single insect, but
never saw any others in this apparatus. I had omitted
to insert within the bulb of the retort a *resting-place*
for these *acari* (they are always destroyed if they fall
back into the fluid from which they have emerged).
It is strange that, in a solution *eminently caustic* and

under an atmosphere of *oxihydrogen gas*, one single *acarus* should have made its appearance.

At first, Crosse mentioned his experiments to only a handful of people, but it wasn't long before the news spread, and in 1837 his name appeared in a local newspaper under the headline 'Extraordinary Experiment'. As might be expected, the report was highly sensationalised (the editor took it upon himself to christen the creatures *Acarus galvanicus*), and stated (untruthfully) that Crosse claimed to have created life from inanimate matter.

The result was quite predictable: Crosse was reviled from one end of the country to the other as a blasphemer who dared to usurp the creative powers of God Himself. He received threats of violence; local farmers blamed him for a blight on their wheat crops; and an exorcism was even performed on the Quantocks. In the midst of the furore, another amateur scientist, one W.H. Weeks of Sandwich, Kent, repeated Crosse's experiments, and also reported the appearance of the *acari*. Weeks' experiments attracted little attention, however, although no less a figure than Sir Michael Faraday became involved at the height of the controversy when he claimed that he had observed similar phenomena during his own experiments (he did not believe, however, that their appearance should be considered as the spontaneous generation of life).

Crosse himself was hurt and confused by the adverse reaction to his researches. In a letter to the writer Harriet Martineau in August 1849, he wrote:

> As to the appearance of the *acari* under long-continued electrical action, I have never in thought, word, or deed, given anyone a right to suppose that I considered them as a creation, or even as a formation, from inorganic matter. To create is to form a something out of a nothing. To annihilate, is to reduce that something to a nothing. Both of these, of course, can only be the attributes of the Almighty. In fact, I can assure you most *sacredly* that I have never dreamed of any theory sufficient to account for their appearance. I confess that I was not a little surprised,

and am so still, and quite as much as I was when the *acari* made their appearance. Again, I have never claimed any merit as attached to these experiments. It was a matter of chance. I was looking for silicious formations, and animal matter appeared instead.

In the face of such vehement criticism of himself and his work, Crosse became yet more reclusive, although he did remarry following the death of his wife Mary in 1846. His new wife, Cornelia, took a great interest in his work and frequently assisted in his experiments.

Andrew Crosse died at Fyne Court on 6 July 1855, in the room in which he had been born. As Rupert T. Gould wrote in his book *Oddities – A Book of Unexplained Facts* (1928):

> For many years he had lived the life of a recluse in his Quantock eyrie, shut off from society, but happy in his marriage and his work. He died as he had lived, an honest man who would make no concession of any kind to popular clamour, but sought truth wherever he might find it. Such men are the true salt of the earth.

What was the true nature of the so-called *acari* that had apparently been generated from Crosse's experiments? Had he really somehow stumbled upon the secret of creating life from inanimate matter? It is now generally agreed that this is unlikely, and that in reality the creatures were either dust- or cheese-mites which had contaminated his carefully prepared apparatus. Although Crosse took every possible precaution to prevent this, he could not have been absolutely certain that his apparatus was completely sterile; for however difficult it may be to create life from lifelessness in an experimental apparatus, it is equally difficult to ensure that the apparatus does not become home to the life that already exists.

7

DEFENDERS OF THE HOLLOW EARTH

THE STRANGE THEORIES OF JOHN CLEVES SYMMES AND CYRUS TEED

Of all the strange and irrational beliefs people have held over the centuries, one of the most bizarre is the idea that our planet is not a sphere floating in the emptiness of space, but rather is a hollow bubble, with everything – people, buildings, continents, oceans and even other planets and stars – existing on the inside. The origin of this curious notion can be traced back to the seventeenth century and the writings of the Jesuit Athanasius Kircher (1602–80), who speculated on conditions beneath the surface of the Earth in a treatise written in 1665 entitled *Mundus Subterraneus* (*The Subterranean World*).

In this work, Kircher draws on the theories and speculations of various medieval geographers concerning the unexplored north and south polar regions. Kircher paid particular attention to the thirteenth-century friar Bartholomew of England, who maintained that 'at the North Pole there is a black rock some thirty-three leagues in circumference, beneath which the ocean flows with incredible speed through four channels into the subpolar regions, and is absorbed by an immense whirlpool'. Having entered this whirlpool, the waters then travel through a myriad 'recesses' and 'channels' inside the planet and finally emerge in the ocean at the South Pole (the continent of Antarctica had yet to be discovered).

Kircher's justification for his ideas was ingenious, if utterly flawed. He claimed that the polar vortices must exist, otherwise the northern and southern oceans would be still and would thus become stagnant, releasing noxious vapours that would prove lethal to life on Earth. In addition, he believed that the movement of water through the body of the Earth was analogous both to the recently discovered circulation of the blood and to the animal digestive system, with elements in sea water extracted for the production of metals and the waste voided at the South Pole. This likening of the Earth to a living entity will doubtless call to mind certain New Age concepts, in particular the so-called 'Gaia Hypothesis'.

The seventeenth-century writer Thomas Burnet (1635?–1715) also suggested that water circulated through the body of the Earth, issuing from an opening at the North Pole. In 1768, this idea was further developed by Alexander Colcott, who added an interesting and portentous twist, that once inside the Earth, the water joined a vast concave ocean – in other words, that the Earth was actually a hollow globe.

In the eighteenth century, the Hollow Earth theory carried considerable intellectual currency; even the illustrious Sir Edmund Halley (1656–1742), discoverer of the comet that bears his name, proposed in the Philosophical Transactions of the Royal Society of 1692 that the Earth was a hollow sphere containing two additional concentric spheres, at the centre of which was a hot core, a kind of central sun. The Swiss mathematician Leonhard Euler (1707–83) concurred and, indeed, went somewhat further, stating that there was a central sun inside the Earth's interior, which provided daylight to a splendid subterranean civilisation.

The apparent credibility of these theories resulted in a brand new sub-genre of fantastic literature. The British writer Joscelyn Godwin provides a brief rundown of the most significant of these tales:

> While medieval theology, as celebrated in Dante's *Divine Comedy*, had found the interior of the earth to be a suitable location for Hell, later writers began to imagine quite the contrary. The universal philosopher Guillaume Postel, in his *Compendium Cosmographicum* (1561), suggested that God had made the Earthly Paradise inaccessible to mankind by stowing it beneath the North Pole. Among the early

novels on the theme of a Utopia beneath the surface of the Earth are the Chevalier de Mouhy's *Lamékis, ou les voyages extraordinaires d'un Egyptien dans la Terre intérieure* (*Lamékis, or the extraordinary voyages of an Egyptian in the inner earth*, 1737), and Ludvig Baron von Holberg's *Nicholas Klim* (1741), the latter much read in Holberg's native Denmark. Giovanni Jacopo Casanova, the adventurer and libertine, also situated Paradise inside the earth. In *Icosameron*, a work supposedly translated by him from the English, he describes the twenty-one years passed by his heroes Edward and Elizabeth among the 'megamicros', the original inhabitants of the 'protocosm' in the interior of our globe. One way into this realm is through the labyrinthine caves near Lake Zirchnitz, a region of Transylvania. The megamicros issue from bottomless wells and assemble in temples, clad in red coats. Their gods are reptiles, with sharp teeth and a magnetic stare.

The literature of the Romantic era, needless to say, is rich in fantasies of polar mysteries and lands within the earth. The best-known works are probably George Sand's *Laura ou le voyage dans le crystal* (*Laura, or the voyage in the Crystal*); Edgar Allen Poe's *The Narrative of Arthur Gordon Pym*; Alexander Dumas's *Isaac Laquédem*; Bulwer Lytton's *The Coming Race*; Jules Verne's *Voyage au centre de la terre* (*Voyage to the Centre of the Earth*) and *Le Sphinx des glaces* (*The Sphinx of the Ice*). Novels by later and less distinguished authors include William Bradshaw's *The Goddess of Atvatabar* (1892), Robert Ames Bennett's *Thyra, a Romance of the Polar Pit* (1901), Willis George Emerson's *The Smoky God* (1908), and the Pellucidarian stories of Edgar Rice Burroughs, creator of Tarzan.

In view of the exciting potential of the Hollow Earth theory, not to mention the literary vogue for such romantic fictions, it was only a matter of time before someone had the bright idea of actually searching for the entrances to the mysterious world apparently lying beneath humanity's feet. Such a man was John Cleves Symmes (1780–1829), who spent a good portion of his life trying to convince the world not only that the Earth was hollow, but that it would be worthwhile to finance an expedition, under his leadership, to find a way inside.

'I DECLARE THE EARTH IS HOLLOW ...'

A native of New Jersey, Symmes enlisted in the United States Army where he distinguished himself for bravery in the French and Indian Wars. Evidently a man of considerable personal integrity, he married a widow named Mary Anne Lockwood in 1808, and ensured that her inheritance from her husband was used to raise her five children (he had five of his own). In 1816 Symmes retired with the rank of Captain and became a trader in St Louis. Two years later, he first announced his beliefs to the world, thus:

> CIRCULAR
>
> Light gives light to discover – ad infinitum
> St Louis, Missouri Territory, North America
> April 10, AD 1818
>
> To all the World:
> I declare the earth is hollow and habitable within; containing a number of solid concentric spheres, one within the other, and that it is open at the poles twelve or sixteen degrees. I pledge my life in support of this truth, and am ready to explore the hollow, if the world will support and aid me in the undertaking.
> Jno. Cleves Symmes
> Of Ohio, late Captain of Infantry
>
> N.B. – I have ready for the press a treatise on the principles of matter, wherein I show proofs of the above positions, account for various phenomena, and disclose Dr Darwin's 'Golden Secret.'
> My terms are the patronage of THIS and the NEW WORLDS.
> I dedicate to my wife and her ten children.

I select Dr S.L. Mitchell, Sir H. Davey, and Baron Alexander von Humboldt as my protectors.

I ask one hundred brave companions, well equipped, to start from Siberia, in the fall season, with reindeer and sleighs, on the ice of the frozen sea; I engage we will find a warm and rich land, stocked with thrifty vegetables and animals, if not men, on reaching one degree northward of latitude 82; we will return in the succeeding spring.

J.C.S.

Of all the academic societies in America and Europe to which Symmes sent his circular, only the French Academy of Sciences in Paris bothered to respond – and that was to say, in effect, that the theory of concentric spheres inside the Earth was nonsense. Undaunted by the total lack of academic interest in his ideas, Symmes spent the next ten years travelling around the United States, giving lectures and trying to raise sufficient funds to strike out for the interior of the planet. He petitioned Congress in 1822 and 1823 to finance his expedition, and even secured 25 votes the second time. Ultimately, the strain of constant travelling and lecturing took its toll on Symmes's health. He died at Hamilton, Ohio, on 29 May 1829. His grave in the Hamilton cemetery is marked by a stone model of the hollow Earth, placed there by his son, Americus.

Symmes's theory of the hollow Earth is described principally in two books: *Symmes's Theory of Concentric Spheres* (1826) by James McBride, and *The Symmes Theory of Concentric Spheres* (1878) by Americus Symmes. (Symmes himself wrote a novel, under the pseudonym 'Captain Adam Seaborn', entitled *Symzonia: A Voyage of Discovery*, published in 1820.)

As we have noted, the Hollow Earth theory attracted the attention of many writers of fiction. Aside from the best-known mentioned above, a number of minor authors explored the topic. In 1871, for instance, Professor William F. Lyon published *The Hollow Globe, or the World's Agitator or Reconciler*, which included many bizarre speculations on open polar seas, the electromagnetic origin of earthquakes and the theory of gravitation (which needed considerable re-working in view of the drastically reduced mass of a hollow planet). The text of the book was apparently received during mediumistic trances by a Dr Sherman and his wife, with Professor Lyon transcribing the material. Among the many curious revelations in this book is the 'great fact that this globe is a hollow or spherical shell with an interior as well as an exterior surface, and that it contains an inner concave as well as outer convex world, and that the inner is accessible by an extensive spirally formed aperture, provided with a deep and commodious channel suited to the purposes of navigation for the largest vessels that float, and that this aperture may be found in the unexplored open Polar Sea'.

The Reverend Dr William F. Warren, President of Boston University, published his book *Paradise Found* in 1885, in which he argued for the origin of the human race at the North Pole. While Warren did not claim that the Earth was hollow, his book nevertheless added to the speculation on the significance of the polar regions, and the idea that the solution to the mystery of humanity's origin might lie there.

In 1896, John Uri Lloyd published his book *Etidorhpa* (the title is 'Aphrodite' reversed). One of the strangest books on the subject, *Etidorhpa* tells the story of one Llewellyn Drury, a Mason and seeker after mystery, who encounters a telepathic humanoid creature without a face. The creature takes Drury into a deep cave in Kentucky, and the two emerge on the inner surface of the Earth, where the adventurer is taught to levitate beneath the rays of the central sun.

A SINGLE BUBBLE IN INFINITE NOTHINGNESS

In 1870, perhaps the strangest of all alternative cosmological theories was formulated by Cyrus Teed: the theory that not only is the Earth hollow, but *we* are the ones living on the inside. Born in 1839 in Delaware County, New York, Teed received a Baptist upbringing. After a spell as a private with the United States Army, he attended the New York Eclectic Medical College in Utica, New York (eclecticism was an alternative form of medicine that relied on herbal treatments). It seems that Teed was greatly troubled by the concept of infinite space, which he could not reconcile with the well-ordered Universe of the Scriptures. While he accepted that the Earth was round (he had little choice, since it had been circumnavigated), he found the notion of a ball of rock floating endlessly through an infinite void so unsettling that he set about attempting to formulate an alternative structure for the observable cosmos.

The answer apparently came to him in a vision in his alchemical laboratory in Utica at midnight one night in 1869. A beautiful woman appeared before him, telling him of the previous lives he had lived, how he was destined to become a messiah, and about the true structure of the Universe. Under the pseudonym Koresh (the Hebrew for Cyrus), Teed published two works: *The Illumination of Koresh: Marvellous Experience of the Great Alchemist at Utica, NY* and *The Cellular Cosmogony*. In his book *Fads and Fallacies in the Name of Science*, Martin Gardner summarises the key points of Teed's outrageous cosmology:

> The entire cosmos, Teed argued, is like an egg. We live on the inner surface of the shell, and inside the hollow are the sun, moon, stars, planets, and comets. What is outside? Absolutely nothing! The inside is all there is. You can't see across it because the atmosphere is too dense. The shell is 100 miles thick and made up of seventeen layers. The inner five are geological strata, under which are five mineral layers, and beneath that, seven metallic ones. A sun at the centre of the open space is invisible, but a reflection of it is seen as our sun. The central sun is half light and half dark. Its rotation causes our illusory sun to rise and set. The moon is a reflection of the earth, and the planets are reflections of 'mercurial discs floating between the laminae of the metallic planes'. The heavenly bodies we see, therefore, are not material, but merely focal points of light, the nature of which Teed worked out in great detail by means of optical laws . . .
>
> The earth, it is true, seems to be convex, but according to Teed, it is all an illusion of optics. If you take the trouble to extend a horizontal line far enough, you will always encounter the earth's upward curvature. Such an experiment was actually carried out in 1897 by the Koreshan Geodetic Staff, on the Gulf Coast of Florida. There are photographs in

later editions of the book showing this distinguished group of bearded scientists at work. Using a set of three double T-squares – Teed calls the device a 'rectilineator' – they extended a straight line for four miles along the coast only to have it plunge finally into the sea [thus proving the Earth to be a concave sphere]. Similar experiments had been conducted the previous year on the surface of the Old Illinois Drainage Canal.

As Gardner observes, Teed was undoubtedly a pseudo-scientist and displayed all the paranoia and obfuscation associated with that infuriating yet fascinating group. His explanations of the structure of the Universe (the ways in which planets and comets are formed, for instance) are couched in impossible-to-understand terms such as 'cruosic force', 'coloric substance' and 'afferent and efferent fluxions of essence'. In addition, he bitterly attacked orthodox science, which sought to impose its erroneous view of reality on a 'credulous public'. According to Gardner, he likened himself '(as does almost every pseudo-scientist) to the great innovators of the past who found it difficult to get their views accepted'.

Teed's scientific pronouncements were combined with apocalyptic religious elements, as demonstrated in the following prophetic announcement:

We are now approaching a great biological conflagration. Thousands of people will dematerialise, through a biological electromagnetic vibration. This will be brought about through the direction of one mind, the only one who has a knowledge of the law of this bio-alchemical transmutation. The change will be accomplished through the formation of a biological battery, the laws of which are known only to one man. This man is Elijah the prophet, ordained of God, the Shepherd of the Gentiles and the central reincarnation of the ages. From this conflagration will spring the sons of God, the biune offspring of the Lord Jesus, the Christ and Son of God.

Unfortunately for Teed, his revelations did not prove of any great interest to the natives of Utica, who took to calling him the 'crazy doctor' and sought their medical advice elsewhere. With his medical practice facing ruin and his wife already having left him, Teed decided to take to the road to spread his curious word. As a travelling orator, he was a spectacular success (he is said to have earned $60,000 in California alone). He was particularly popular in Chicago, where he settled in 1886 and founded the first College of Life and later Koreshan Unity, a small communal society.

In the 1890s, Teed bought a small piece of land just south of Fort Meyers, Florida, and built a town called Estero. He referred to the town as 'the New Jerusalem', predicted that it would become the capital of the world, and told his followers to expect the arrival of eight million believers. The actual number who arrived was something of a disappointment, being closer to 200; nevertheless, the happy, efficient and hard-working community seems to have functioned extremely well. Their strange ideas notwithstanding, the members, male and female alike, were treated as equals, which is no bad thing.

Teed died in 1908 after being beaten by the Marshal of Fort Meyers. He had claimed that after his death he would be taken up into Heaven with his followers. They dutifully held a prayer vigil over his body, awaiting the event that, unsurprisingly, did not take place. As Teed's body started to decompose, the county health officer and ordered his burial. He was finally interred in a concrete tomb on an island off the Gulf Coast. In 1921 a hurricane swept the tomb away; Teed's body was never found.

8

THE UNIVERSE NEXT DOOR

THE POSSIBILITY OF PARALLEL WORLDS

E. E. cummings wrote that there's a hell of a good Universe next door. But was he right?

Our Universe (at least, the part of it we can observe) is approximately 14,000 million light years in radius, and contains hundreds of thousands of millions of galaxies. It seems like the craziest of notions to suggest that this unimaginable immensity is but one tiny bubble in an infinitely larger reality – a 'multiverse'. And yet that's precisely what many cosmologists have come to believe.

The Swedish physicist Max Tegmark has come up with an intriguing scenario in which an elderly man conducts an experiment with the aid of an extremely nervous assistant. The elderly man has set up a machine gun in the centre of a laboratory, its firing mechanism connected to a control panel at which his assistant stands. The firing mechanism has been modified so that, when the assistant presses the button, the weapon will either fire or not fire. Whether it fires or not is determined completely randomly.

The man stands in front of the machine gun and reassures his young assistant, telling her that he's old: he has nothing to lose if the experiment fails. Taking a deep, shuddering breath, she presses the firing button on the control panel.

There is a dull click: the machine gun has not fired. She presses the button again. Again, the weapon clicks harmlessly. She presses the button again, and from her point of view, the machine gun fires, peppering the old man with bullets and killing him. But ... from the old man's point of view, the weapon merely clicks once again. He is still alive. His assistant pushes the firing button again and again, and every time, the weapon fails to fire. No matter how many times his assistant pushes the button, from the old man's point of view, the machine gun fails to fire. After she pushes the button for the hundredth time, the man steps out of the gun's firing line, and proclaims triumphantly that he was right: he cannot die.

How could this be? Surely, either the gun fires or it doesn't. Either the old man survives, or he is killed. How can he both survive and be killed? This scenario only makes sense if we are living not in a single, unique Universe, but in a *multiverse* composed of countless universes. In other words, each time the assistant pushes the firing button, the Universe splits into two entirely separate realities – one in which the gun fires and kills the man, and one in which it doesn't. The next time she pushes the button, the Universe splits again into two more separate realities, and so on, and so on.

In some realities, the gun fires and the old man dies; but there will always be other realities in which it doesn't fire, and he survives. According to Max Tegmark, the man would have no awareness of the realities in which the machine gun fired, since he would be dead. 'The only realities he would continue to be aware of would be the ones in which he survived. It was inevitable therefore that after fifty clicks, a hundred, two hundred, he would step out of the firing line of the machine gun having discovered he was immortal.'

How did the bizarre notion of the multiverse arise? Doesn't our single Universe contain enough wonders and mysteries to be going on with? The fact is, it is the outlandish nature of some of those wonders and mysteries that has led physicists and cosmologists to speculate that there *must* be an infinite number of other universes, in order for this one to make sense.

There is the problem of the so-called 'fine-tuning' of the Universe, for instance. For most people going about their daily lives, the Universe is the way it is because . . . well, it just *is*. There are planets and stars and galaxies; there are atoms and molecules and liquids and gases, and everything seems to work just fine. We take for granted the fact that the laws of physics are the way they are: after all, most of us have more pressing concerns, like paying bills and putting food on our tables. It's of little concern to us why gravity is the strength it is, or why the other fundamental forces of nature are the way they are.

But if we take the time to ask *why*, as physicists and cosmologists have done, we will find the implications truly staggering. It was the British astronomer Sir Fred Hoyle who first noticed in the 1950s that the Universe appears to be 'fine-tuned' so that life and consciousness can exist. We have been told a number of times, in popular books and television programmes, that we are made of 'star-stuff': that the atoms in our bodies were forged in the nuclear furnaces at the centres of stars. Heavy atoms are built up from lighter ones, and Hoyle discovered the processes by which this occurs. These processes, however, depend on a number of strange coincidences. Only if the nuclei of three atoms (beryllium-8, carbon-12 and oxygen-16) have just the right energy can hydrogen, the lightest and simplest atom, be assembled into the heavier atoms which are essential for the development of life.

Hoyle's example of fine-tuning isn't the only one. Says Max Tegmark: 'Many instances have been found in which if a certain fundamental force of nature were slightly weaker or stronger, or if a certain fundamental particle were slightly lighter or heavier, there would be no galaxies or stars or planets, and hence no human beings.'

We've already mentioned gravity. It seems such a mundane thing. It keeps us on the surface of the Earth, and keeps the planets orbiting the Sun, and we're not above cursing it when we drop something valuable (or wet) on the floor. But if gravity were not precisely the strength it is, we would not exist. If the force of gravity were just a few per cent weaker than it is, then the gas from which stars are made would not have been able to contract to the point where nuclear fusion could commence. There would be no stars, and hence no life. And if gravity were only a few per cent stronger than it is, it would increase the temperature in the cores of stars, causing them to burn their fuel in a few million years instead of the billions required for life in their planetary systems to develop.

In addition to gravity, there are the strong and weak nuclear forces, each of which has a direct bearing on the way our Universe developed. If either of these forces were stronger or weaker, we would not be here. Physicists like Tegmark maintain that the fine-tuning of nature cannot be dismissed as mere coincidence. 'There are only two possible explanations,' he says. 'Either the Universe was designed specifically for us by a Creator. Or there exists a large number of universes, each with different values of the fundamental constants, and not surprisingly we find ourselves in one in which the constants have just the right values to permit galaxies, stars and life.'

There is further evidence for multiple universes in quantum theory, the most successful scientific theory ever devised. Quantum theory describes the behaviour of the basic building blocks of matter: atoms and their components. As more than one physicist has said, quantum theory has made the modern world possible: it has given us everything from microwave ovens to space shuttles.

It also tells us that elementary particles like electrons and photons can be in several places at once. And, according to the great physicist Hugh Everett III (1930–82), this outrageous property is not confined to the microscopic world, but also applies to the macroscopic world of tables and trees, of houses and cars and people. In short, everything that you see around you, the chair on which you are sitting, the walls of the building around you, the town or city outside your window, *all simultaneously exist in several places at once.*

Everett's idea has come to be known as the Many Worlds interpretation. According to Tegmark: 'If you observe a table which is in two places at once, your mind will also end up in two states at once – one state which perceives the table in one place and one which perceives it in another place!'

How can this bizarre notion be an accurate reflection of reality? It seems to be completely counter-intuitive, to go against our concept of 'common sense'. The answer may lie in the so-called 'quantum fluctuation' which cosmologists believe gave rise to the Big Bang in which the Universe was created. According to quantum mechanics, matter and energy can appear spontaneously out of the vacuum of space (which is actually not a vacuum at all, but is seething with energy) in a kind of 'cosmic hiccup'. And if a quantum fluctuation can happen once, it can happen again and again, resulting in the birth of countless universes. According to the so-called 'anthropic principle', these universes will have different physical laws, some of which will be unsuitable for the development of life and intelligence, and some of which will.

Some cosmologists are against the idea of multiple universes, for the straightforward reason that it is not (at least at present) a testable hypothesis. If such other universes exist, how could we possibly detect them? As we noted at the beginning of this chapter, our Universe is about 14,000 million light years in radius – but that's just the part we can *observe*; in other words, that's as far as light has had time to travel during the approximately 14,000 million years of the Universe's history. There may be (in fact, there probably *is*) a whole lot more of the *physical* Universe out there that we simply can't observe because the light hasn't had time to reach us yet. And if we can't even observe the totality of our own Universe, it's even more difficult (impossible, indeed) to observe another universe entirely. According to some astronomers, when you start talking about other universes, you're leaving the realm of science behind and entering the realm of metaphysics.

Other cosmologists have a problem with the whole idea of the fine-tuning of the Universe which allows our kind of life to exist. According to cosmologist Andreas Albrecht of the University of California at Davis, the whole idea of fine-tuning ignores the vast potential for other forms of life in this and other universes. The proponents of the anthropic principle, he says, 'don't know what it takes to have life in the universe. There could be forms of life out there that we haven't even thought of!'

If other universes exist, could we ever detect them? Some theorists speculate that it might be possible to detect gravitational energy 'leaking' into our Universe from others; and it has even been suggested that the reason why gravity in our Universe is such a weak force is that it is leaking through from a universe in which it is much stronger.

This sounds like a pretty wild idea; but Albrecht reminds us that many theories that were once considered fanciful have turned out to be true. Quoting Albrecht in an article on the website Space.com, science journalist Andrew Chaikin writes that in the late nineteenth century, 'most scientists didn't accept the idea that matter was composed of atoms – an idea supported not by direct observation, but by inferences based on theories of temperature, heat and viscosity'. He quotes:

> 'Atomic theory had some great things to say about that, and seemed to give a consistent, unified picture,' Albrecht says, but 'the majority of physicists at that time didn't really believe atoms existed; they thought it was just some flight of fancy'.

Is the idea of a multiplicity of universes another 'flight of fancy' that will one day turn out to be true? Only time, perhaps, will tell.

9

THE OMEGA POINT THEORY

SCIENTIFIC PROOF OF THE AFTERLIFE?

Imagine yourself at the end of your life, taking your turn at last on the great frontier of night, as every human being must. Yours may have been a happy life, filled with many friendships and loves and achievements. It may have been hard, filled with adversity and suffused with loneliness and pain. Or it may have been quite nondescript, with little to distinguish it from the equally mundane lives of the teeming millions around you. For better or worse, yours has been a mayfly existence, a flicker of mentation in the staggering immensity of the cosmos.

If you are lucky enough to approach the end in comfortable surroundings with which you are familiar, surrounded by family and friends whom you love, and who you know will mourn your passing, you may ponder all the things you didn't manage to do. For even the tiny Earth is vast to the tinier creatures who swarm across its surface; and even the fullest life cannot encompass all that there is to experience in this little world of ours. You may think of the places you should have visited, the marvellous sights you should have seen, the things you could have achieved, but for which you never quite found the time.

You feel the end approach: the ultimate mystery awaits you, and you feel fear impinge upon the tranquillity and resignation you had experienced up until this moment. What will death bring? The annihilation of your awareness for all time . . . or something else, something much, much more? You close your eyes, and breathe your last breath on Earth . . .

. . . and awaken in a comfortable room – perhaps one you remember from childhood – suffused with a soft light whose source you cannot see, which seems to come from the very air itself. You are quite alone . . . and yet, somehow, you *know* exactly where you are without anyone telling you. It's as if knowledge is flooding into your mind from some unknown source. You realise that death does not bring annihilation; your awareness has not plunged into the insensate void of lightless eternity, but continues . . . here.

You look around the room, which you know has been chosen by the Intelligence controlling this place to be the least frightening to you upon your awakening. Perhaps you were always ready to embrace new concepts while alive: the unknown did not frighten you; and for that reason the Intelligence allows you to know what is outside the world into which you have awoken.

Outside, the temperature is trillions of degrees Celsius; the pressure would crush planets, stars, galaxies out of existence. You have awoken at the final instant of space and time, the last heartbeat of the Universe before it collapses into the singularity from which it sprang countless aeons ago. And yet eternity lies before you. You are an *emulation* of what you were, preserved in the computer at the End of Time. You have been resurrected inside the Omega Point.

THE AFTERLIFE AT THE END OF TIME

This is the possibility envisaged by the American physicist Frank Tipler, which he explains in his astonishing book *The Physics of Immortality: Modern Cosmology, God and the Resurrection of the Dead*. The Omega Point theory states that in the far future the entire Universe will be transformed into a single all-powerful, infinitely intelligent computer. This Intelligence will choose to resurrect every being that has ever lived, in the form of computer emulations, which it will place in an immensely sophisticated virtual reality environment. The Omega Point, argues Tipler, will be indistinguishable from what we consider to be God; and the environment it will create will be indistinguishable from our concept of Heaven.

Although Tipler first began thinking about the Omega Point as a postdoc in Berkeley in the 1970s, it wasn't until he collaborated with the British physicist John Barrow on a book called *The Anthropic Cosmological Principle*, published in 1986, that he began to think of the Omega Point as more than an entertaining intellectual game. In *The Anthropic Cosmological Principle*, Tipler and Barrow considered what might happen if intelligent machines converted the entire Universe into a gigantic information-processing device. They concluded that in a closed universe (that is, one which stops expanding and collapses back on itself) information-processing capacity would approach infinity in the final stages before the final singularity was reached.

The term 'Omega Point' was originally used by the Jesuit mystic and scientist Pierre Teilhard de Chardin, who had speculated on a distant future in which all living things merged into a single entity, a divine being embodying the spirit of Christ. When Tipler read an essay by the German theologian Wolfhart Pannenberg proposing that future humans would live again in the mind of God, he realised that the most troubling question regarding the Omega Point (what would an infinitely intelligent being actually *do*?) had just been answered. The Omega Point would have the power to re-create everyone who had ever lived. How would it do this? By brute-force resurrection, says Tipler: it would take advantage of something called the Beckenstein Bound, which places an upper limit on the number of states in which a system can exist. Since the Beckenstein Bound also applies to human beings, the Omega Point will simply re-create every state of every human being who has ever lived.

Of course, it will not re-create the people in the physical form in which they exist today: as noted earlier, the conditions near the End of Time would annihilate a human body instantly. Instead, we will be resurrected as emulations inside the Omega Point computer. It is vital to recognise the distinction Tipler makes between computer *simulations* and computer *emulations*. A simulation of an object or person is simply that: a *representation* of the object or person. An emulation is an *exact* duplicate, indistinguishable from the original – right down to the positions of the fundamental particles of which it is composed. In short, a computer emulation of a person *would be that person.* Hence, from the point of view of the person in the little scenario described at the beginning of this chapter, death is followed instantaneously by a resuming of consciousness inside the Omega Point computer at the End of Time (although billions or trillions of years may have elapsed in the meantime).

As if this idea were not bizarre enough, Tipler adds another level of strangeness, asserting that the Omega Point actually created the Universe – even though the Omega Point itself has yet to come into existence! Tipler attempts to resolve this apparent paradox by explaining that the future should be our frame of reference, since it dominates our cosmic history. This is, in effect, a reiteration of the Argument from Design to explain the origin of the Universe. Says Tipler: 'We look at the universe as going from past to future. But that's our point of view. There's no reason why the *universe* should look at things that way.'

Tipler writes in *The Physics of Immortality*:

> Some scholars have argued that this view of God, that He/She is to be considered primarily a future being, was already present in the very beginning of ancient Judaism. When God spoke to Moses out of the burning bush, Moses asked Him for His name. According to the King James translation of the Bible, God replied 'I AM THAT I AM . . . say unto the children of Israel that I AM hath sent me [Moses] unto you' (Exodus 3:14). However, in the original Hebrew God's reply was '*Ehyeh Asher Ehyeh*.' In Hebrew, the word *Ehyeh* is the future tense of the word *haya*, which means 'to be.' That is, God's reply to Moses should be translated 'I WILL BE WHAT I WILL BE . . . Tell the children of Israel that I WILL BE sent me to you.' . . . The Jewish German philosopher Ernst Bloch and the Catholic German theologian Hans Küng have both pointed out this true future tense translation, and emphasize that the God of Moses should be regarded as an 'End- and Omega-God.' The Omega Point God described in [*The Physics of Immortality*] is definitely a God Who exists mainly at the end of time.

What would we do in the Heaven of the Omega Point? For, as many thinkers have noted, the biggest problem about eternity is that there is an infinite amount of time available in which to become bored. Tipler suggests that the information-processing capacity of the Omega Point computer would be so vast (infinite, indeed) that to recreate every version of the Universe that is logically possible would be an insignificant use of its resources. We could explore this and all other universes; we could even design our own universes, which the Omega Point would then 'realise' as virtual realities.

And yet, even trillions of years spent exploring and designing universes shrink to total insignificance compared with eternity. For this reason, Tipler believes that the Omega Point would offer us the option of joining with it. We would become as gods – superbeings capable of enjoying new experiences for the rest of eternity.

The Omega Point would be able to exist for eternity, even though the dying Universe which gives rise to it would come to an end in the final singularity. How can this be? The reason is that the computers of the Omega Point would process information at an infinite speed: in other words, they would be running *faster than time runs out*. Thus, if it were possible to observe the final moments of the Universe from the 'outside', as it were, an observer would see it collapse into nothing in the final singularity (sometimes known as the Big Crunch); however, from the point of view of the Omega Point, and the beings who have been resurrected inside it, their perception of time would diverge into eternity. There would be *no End of Time*.

The idea of an infinitely fast, infinitely powerful computer is an incredible one. Could such a device ever be developed?

UNLIMITED COMPUTER POWER?

It's a familiar nightmare – a science fiction cliché, in fact – and one that will be familiar to anyone who has seen the *Terminator* films. At some time in the not-too-distant future, computers will achieve self-awareness and determine that humanity has made a complete hash of running the planet, and decide to have a go themselves. This scenario (which is almost always presented as something deeply unpalatable and to be avoided at all costs) received a slightly more sophisticated treatment in the *Matrix* films, in which a powerful machine intelligence uses human beings as batteries while confining their minds in a virtual reality world.

This idea makes for an enjoyable bit of cinematic hokum, but is it something we should be genuinely worried about? Will computers ever achieve true intelligence? And if they do, will they ever slip the bonds of human control and replace us as the dominant life form on this planet?

Anyone who has wrestled with a recalcitrant computer will think this scenario unlikely in the extreme. It's hard to imagine how machines that shut down 'improperly' on a whim, and then consider it necessary to inform us of the fact when we manage to restart them, could present a serious threat to the continued supremacy of human beings.

And yet, in spite of the myriad frustrations they cause us every day, there is no doubt that computers have transformed the way we live. They perform every function, from controlling the cycle in your washing machine to predicting the weather, from monitoring the heartbeat of a premature baby to telling the time in Acapulco. It's hard to see what else they could do, given the degree to which we have become dependent on them.

But there is a device that puts to shame even the most powerful IBM or Cray supercomputer, and it is a device which every one of us possesses. It is the few pounds of soft tissue nestling behind your eyes: the human brain, whose processing capacity dwarfs that of the most sophisticated computers.

So how powerful is the human brain? What is its processing capacity? The units for measuring the speed at which a computer processes information are known as 'flops', which stands for *fl*oating *p*oint *op*erations per second. A floating point operation is simply the addition, subtraction, multiplication or division of two numbers expressed in scientific notation. The kind of computer you have in your home is pretty fast: it can do a few megaflops (one million floating point operations per second). In the mid-1980s, the fastest supercomputer available, the Cray-2, had a speed of 1 gigaflop (or one billion flops). Only half a decade later, the speed of the fastest supercomputer reached 10 gigaflops.

In *The Physics of Immortality*, Tipler writes:

In January 1992, Thinking Machines Inc. shipped a 100-gigaflop machine, the CM-5, to Los Alamos research labs. The cost of this machine was $10 million, a standard price for a state-of-the-art supercomputer. Danny Hillis, the chief scientist for Thinking Machines, announced at the time that his company was ready to build a 2-teraflop (that's two *trillion* flops) any time someone would come up with the money to pay the $200 million price. (A teraflop computer is sometimes called an *ultracomputer*.)

Tipler goes on to ask how rapidly the human brain processes information in comparison. The answer is stunning.

About 1% to 10% of the brain's neurons are firing at any one time, at a rate of about 100 times per second. If each neuron firing is equivalent to a flop, the lower number gives 10 gigaflops. If each synapse is equivalent to a flop at each firing, then the higher number gives 10 teraflops. [The computer scientist] Jacob Schwartz estimates 10 million flops as an upper bound to the amount of power required to simulate a single neuron. If this is the actual requirement, then 100,000 teraflops would be required to simulate the entire brain.

Tipler thinks that if the rate at which computer power increases is maintained ('computer speeds have increased over the past forty years by a factor of 1,000 every twenty years'), then we can expect to see personal computers with human-level information processing capacity by the year 2030.

But will the ultracomputers of the future actually be able to *think*? Will they possess a consciousness comparable to human consciousness? The only way to be sure would be to talk to the computer, to put it through what is known as the Turing Test, named after the great British computer scientist Alan Turing, who proposed it in the 1950s. In the Turing Test, a computer is placed in one room, and a human being in another. There is a computer screen and keyboard outside the two rooms, connected to another screen and keyboard in the room containing the human being, and connected directly to the computer in the other room. The person outside does not know which room contains the human being, and which contains the computer. Then the test begins, with the outside observer typing questions on his keyboard and analysing the replies.

If, after days or weeks or years of typing in messages and receiving answers, the observer cannot tell which room contains the human being and which contains the computer, then the computer will have passed the Turing Test: it will have demonstrated its ability to think exactly like a human being.

We have yet to build a computer which possesses human-like consciousness, still less one with *superhuman* intelligence. But that day may be closer than we think. In the previous chapter, we discussed the Many Worlds interpretation of quantum mechanics, in which countless alternate universes are constantly branching off from our own. We may be able to utilise these parallel realities in what are known as 'quantum computers'. A quantum computer would exploit the ability of fundamental particles to be in many places at once, to do many calculations at once.

Although quantum computers have been built, the field is very much in its infancy, and these prototypes are only capable of processing a handful of binary digits ('bits'). However, we can confidently predict that within a few decades much more powerful quantum computers will be built – and even today's ultracomputers will pale into insignificance compared to them. Says the science writer Julian Brown: 'If you imagine the difference between an abacus and the world's fastest supercomputer, you would still not have the barest inkling of how much more powerful a quantum computer could be compared with the computers we have today.'

It is entirely possible that a quantum computer might be built which could carry out more calculations at any one time than there are fundamental particles in the Universe. The Oxford physicist David Deutsch asks an important question: if a quantum computer performs more calculations in any instant than there are fundamental particles in the Universe, then *where* are these calculations being carried out? The Universe does not possess the physical resources to do what the computer is doing.

The answer lies in the Many Worlds. A quantum computer does not have to rely on a single universe to carry out its calculations: different parts of the calculations are performed in different realities. In effect, quantum computers achieve their incredible feats by utilising huge numbers of versions of themselves in other neighbouring realities.

The unprecedented rate at which quantum computers are able to process information may eventually lead to the construction of truly intelligent, conscious machines. And such would be their intelligence that they would be able to design yet *more* intelligent machines, and so on. Human-level intelligence would quickly be left far behind – unless we decided to enter what is known as the 'post-human' phase of our existence, and to merge with these super-intelligent entities. At which point, we would have taken a significant step towards the Omega Point.

10

ALIEN ARCHAEOLOGY

ANOMALOUS STRUCTURES ON MARS AND THE MOON

In 1959 the United States National Aeronautics and Space Administration (NASA) commissioned the highly respected Brookings Institute in Washington, DC to conduct a study which was entitled 'Proposed Studies on the Implications of Peaceful Space Activities for Human Affairs'. At that time, NASA was still in its infancy, having been created by Congress only a year before, in the 1958 Space Act, so it was entirely logical that such concerns should be addressed. Experts in a wide range of disciplines from all over the United States participated in the study, which took a year to complete.

The Brookings Institute Report contains a section entitled 'Implications of a Discovery of Extraterrestrial Life'. In part, it concluded that 'cosmologists and astronomers think it very likely that there is intelligent life in many other solar systems', and that 'artefacts left at some point in time by these life forms might possibly be discovered through our space activities on the Moon, Mars, or Venus'.

The section on extraterrestrial implications dealt with 'the need to investigate the possible social consequences of an extraterrestrial discovery and to consider whether such a discovery should be kept from the public in order to avoid political change and a possible "devastating" effect on scientists themselves – due to the discovery that many of their own cherished theories could be at risk'.

The report concluded that we should not be told; it concluded that the risk to our civilisation would be too great, should alien artefacts ever be discovered in the course of our exploration of space. The main reason given was the unfortunate history of our own species, in which many cultures have been devastated by contact with their technological superiors.

A MARTIAN COMPLEX?

In recent years, there has been a great deal of controversy regarding certain photographs taken by American and Russian space probes. Anyone with the slightest interest in the subject of alien contact (and many with none whatsoever) will be aware of the so-called 'Face on Mars', also known as the 'Martian Sphinx', a 1.6-kilometre-long mesa first photographed by one of the two Viking probes as it passed over the region of Cydonia in 1976. In addition to the Face, the Viking images revealed several other unusual surface features, including several pyramid-like structures and the 'Tholus', a circular mound featuring what looks like a spiral ramp.

When the Viking images were first released by NASA, interest briefly centred on frame 35A72, which showed a landform resembling a humanoid face. NASA was quick to dismiss this interpretation, insisting that the curious image of the Face was nothing more than a 'trick of light and shadow'. However, Vincent DiPietro and Gregory Molenaar, engineers at the Goddard Space Flight Centre and computer imaging experts, were so impressed that they searched the other 60,000 Viking frames until they found another photograph of the same feature, but taken 35 orbits later and with the Sun at a different angle. With the aid of various computer enhancement techniques, they were able to increase the resolution of the Face. The results were stunning. It seemed that the Face had eye sockets complete with eyeballs, a mouth with discernible teeth, and regularly spaced striations that appeared to be a kind of headdress.

Since the orientation of the Face was directly upwards, another researcher, Richard Hoagland, concluded that if it had been intelligently constructed, its function must have been to draw the attention of someone looking down from a great height, perhaps even from orbit. An examination of the surrounding landscape has yielded even more amazing results: the presence of yet more apparently artificial structures, including a colossal five-sided pyramid (now known as the DiPietro Molenaar Pyramid), and the Tholus. In fact, an observer standing at the centre of what Hoagland calls 'the City' would have a fine view of the Face in profile across the Martian desert.

Hoagland, who has become the best-known and most vocal of 'Martian anomaly' researchers, suggests that not only are the structures artificial, but some may be hollow, the upper sections of some structures having collapsed in places, raising the possibility that they were once living quarters. In support of this theory, he cites the environmental conditions on Mars, for example the very low atmospheric pressure (about a hundredth that at sea level on Earth). In order for any complex life form to live in such a place, he reasons, some sort of artificial environment would have been essential.

This has led Hoagland to speculate that the structures might be comparable to the 'arcologies' (architectural ecologies) proposed by the architect Paolo Soleri in the 1960s. These colossal buildings would be completely self-contained, and capable of supporting millions of inhabitants. They would contain everything needed by a technological society, including factories, greenhouses, living and recreation areas and energy-generating equipment. (A small-scale example of this kind of self-contained community was the experimental Biosphere II in Arizona – 'Biosphere I' being the Earth itself.)

The Martian structures are somewhat larger than Biosphere II, however: some of the 'pyramids' are 1.5 to 3.5 kilometres across. Their vast size suggests that, if they are artificial (and that is a very big 'if'), they may have housed millions of individuals, perhaps the last members of a dying race. (Devotees of fantastic fiction may detect a similarity here with the Last Redoubt, the gigantic metal pyramid which is the final home of a far-future humanity in William Hope Hodgson's extraordinary fantasy *The Night Land*, first published in 1912.)

Hoagland also claims to have discovered a complex geometric pattern linking many of the structures, one that further reinforces the notion that they were built with a definite purpose in mind. According to Hoagland, the geometry expresses two fundamental constants of nature: *pi* (the ratio of the circumference of a circle to its diameter) and *e* (the base of natural logarithms). *Pi* divided into *e* gives the ratio 0.865, a ratio that is present throughout the 'Cydonia complex'. Hoagland states that the mathematical data encoded in the complex confirm certain predictions made by astrophysicists, in particular that spinning objects such as stars and planets should display energy upwellings at specific latitudes (19.5° north or south). This angle is also repeated throughout the Cydonia complex.

Hoagland then discovered a pattern ranging throughout the Solar System, based on latitude 19.5°. At this latitude on Earth lies the Hawaiian chain of volcanoes; on Mars we find the gigantic shield volcano Olympus Mons, the largest mountain in the Solar System; and on Jupiter there is the Great Red Spot, a storm system larger than the Earth. Hoagland then predicted that the Voyager probe would reveal a spot on Neptune at latitude 19.5°, two weeks before the spacecraft arrived at the planet. He was proved correct.

According to Hoagland, these discoveries represent the existence of a 'new physics', based on hyperdimensional mathematics. As he reminds us, energy flows 'downhill' (from hot to cold, from higher to lower levels). He maintains that a spinning object, such as a star or planet, will exhibit energy anomalies as a result of its connection to higher and lower dimensions. The implication here is that whoever built the Cydonia complex was well aware of hyperdimensional physics, and was capable of utilising the 'free energy' arriving from higher dimensions.

AN UNFORTUNATE TRUTH

Ever since it was first photographed in the 1970s, the Face on Mars has captured the imaginations not only of believers in UFOs and alien life, but also of members of the general public who had never before given any serious thought to these topics. It featured in an episode of the phenomenally successful TV series *The X-Files*, and was also the inspiration for the film *Mission to Mars*, starring Tim Robbins and Gary Sinise. Countless books and magazine articles have cited it as indisputable proof that there was once a sophisticated civilisation on Mars, almost certainly long-vanished, but perhaps – just *perhaps* – still living beneath the cold dead surface of the red planet.

Needless to say, NASA and the astronomy community were

much more cautious, if not outright dismissive of these claims, maintaining that the giant mesa only gave the *impression* of humanoid facial features. It was, they said, no more than a trick of light and shadow, combined with the low resolution of the Viking images. Were it to be re-imaged with more sensitive equipment, it would doubtless turn out to be simply an oddly shaped lump of rock.

One can sympathise with NASA's position. If anything could have secured the space agency unlimited funding for the foreseeable future, it would have been unequivocal proof of an extraterrestrial civilisation on our planetary next door neighbour. One supposes that it must have been very tempting to come right out and say: 'Yes, the Face is artificial, now let's get out there and do some exploring!' One can also imagine the US Government responding: 'No problem! A hundred billion dollars to get started? Will you take a cheque?'

But that's not the way scientists work: they have their reputations to consider, and it's much easier and cheaper to send unmanned probes to check out astonishing possibilities. That's exactly what NASA did with their Mars Global Surveyor probe, which arrived at Mars in September 1997. One of the mission's top priorities was to take some more photographs of the Face. As the chief scientist for NASA's Mars Exploration Program, Jim Garvin, said: 'We felt this was important to tax-payers . . . we photographed the Face as soon as we could get a good shot at it.'

On 5 April 1998, the Global Surveyor flew over the Cydonia region, and its Mars Orbiter Camera started taking pictures ten times sharper than those taken by Viking 18 years earlier. This was the day Richard Hoagland and the many other believers in Martian artefacts had been waiting for; the day when the Face would be revealed in all its weird glory: proof at last that humanity was not alone in the Universe.

The images were posted first on the NASA Jet Propulsion Laboratory (JPL) website, and they revealed . . . nothing of particular interest. What had appeared to be an unmistakably humanoid face in the original Viking images was, in fact, a natural landform containing no recognisable facial features whatsoever. The Viking images had indeed been tricks of light, shadow and low resolution.

It's very difficult to watch one's theories or beliefs torn to shreds in front of one's eyes, and so, quite understandably, criticism of the new images began almost at once. The location of the Face (at 41° north Martian latitude) was experiencing atmospheric cloud and haze during April 1998, and it was suggested that the true appearance of the Face might have been distorted.

The NASA mission controllers were open to the idea of photographing the Face again, although it would be another three years before the Global Surveyor was again in the correct position to do so.

On 8 April 2001, the spacecraft flew once again directly over the Face. The atmospheric conditions were perfect: there wasn't a cloud in the Martian sky, and the probe had a completely unobstructed view of the landform below. According to Garvin: 'We had to roll the spacecraft 25 degrees to centre the Face in the field of view.' The camera was operating at maximum resolution, with each pixel in the resulting image covering just 1.56 metres – about 27 times better than the original Viking images. 'As a rule of thumb,' said Garvin, 'you can discern things in a digital image three times bigger than the pixel size. So, if there were objects in this picture like airplanes on the ground or Egyptian-style pyramids or even small shacks, you could see what they were.'

But that's not what the picture showed. It was a clearer version of the images captured in 1998, of an entirely natural-looking butte or mesa, no different from the many other such landforms scattered across the Cydonia region.

Notwithstanding this disappointing conclusion to the saga of the Face, the mesas of Cydonia are of great interest to planetary geologists because the region lies in a transition zone between the heavily cratered highlands to the south, and the smooth lowland plains to the north. In other words, the Cydonian mesas may well once have lain on the shores of an ancient ocean.

An amateur rock climber, Jim Garvin imagines what it would be like for future astronauts to climb the Face and look out across the Martian landscape: 'From there the view would be spectacular. To the south the ground would slope upwards, toward the highlands. To the north the terrain would descend toward the plains. Looking around you would see a barren landscape dotted with buttes, mesas, and impact craters.'

Perhaps one day, explorers will do just that. But if they are also to search for the relics of an alien civilisation, they will have to look elsewhere than the 'Face on Mars'.

THE GREAT GALACTIC GHOUL

Somewhere out in the depths of interplanetary space between Earth and Mars, *something* lurks: a dangerous force that has damaged or destroyed more than 20 of our probes to the red planet. At least, that's the rueful joke at NASA, where the 'force' has become known as the Great Galactic Ghoul, a term coined by the *Time* magazine journalist Donald Neff.

The Americans aren't the only ones to have fallen foul of the Ghoul. Towards the end of 2003, Japan announced the failure of its own Mars probe, named Nozomi, which means 'hope' in Japanese, and which came to grief without even reaching its goal. Said Louis Friedman, executive director of the Planetary Society, a space exploration advocacy group based in Pasadena, California: 'Space is very unforgiving. You can do 10,000 things right, but do one thing wrong, and you are doomed.'

Nozomi was merely the latest casualty in a long line of heroic failures in the history of humanity's quest to explore the red planet. The problems began almost as soon as the ship blasted off on 4 July 1998. First, it veered off course and had to use precious fuel to correct its flight path, which in turn required the mission controllers to recalibrate its trajectory. In April 2002, the ship was caught in a solar flare and lost power. Hundreds of attempts were made to switch the power back on, as the craft tried to tap the Sun's energy to jump-start its frozen systems. Although this audacious plan worked, the Nozomi was finally defeated by a simple electrical short in its navigation system, which caused it to miss Mars by a hair's breadth (in cosmic terms), and head off irretrievably into deep space, to be lost for ever.

That doom also befell NASA's Mars Observer spacecraft, which was launched from Florida on 25 September 1992. After a flight lasting nearly a year, the one-billion-dollar Mars Observer reached its destination on 21 August 1993. Up to this point, everything had gone according to plan, and the mission controllers were confident that the probe would successfully insert itself into a near-polar orbit from which it would conduct a detailed photo-reconnaissance of the planet.

There was just one procedure that needed to be completed before the Observer could begin its task of photographing the Martian surface and taking sensor measurements of the thin wisp of atmosphere shrouding the mysterious planet.

This procedure necessitated turning off the radio so that its delicate components would be protected during the pressurisation of the craft's fuel tanks: in order to achieve its polar orbit, the Observer would need to fire its thrusters – always a potentially risky operation. The transmitter was duly switched off, the tanks were pressurised and the thrusters fired. However, when the time came for the radio transmitter to be reactivated, the Observer remained ominously silent.

At this point, the mission controllers were not unduly alarmed: the craft contained a back-up transmitter which would take over in the event of damage or malfunction. An instruction was sent to the Observer to switch on its additional transmitter. Again, there was only silence.

Now the controllers realised that they had a potentially serious problem on their hands. If both transmitters were out of action, the probe would be useless; even if its other instruments were still functioning and able to collect data, it would have no way of sending the data back to Earth.

There was just one possibility remaining, one chance that the mission could be saved. Perhaps the Observer was still operational, but had suffered damage to its receiver. In the event of loss of contact with Earth, the spacecraft was programmed to wait for five days, and then send a signal requesting instructions.

So the mission controllers waited . . . and waited. The five days passed, and still no signal was received from the Mars Observer, tens of millions of miles away in the depths of space. The probe was lost, never to be heard from again.

On the other side of the world, the Russians looked on with sympathy, for they had experienced more than their fair share of disasters in near-Mars space. In fact, in their 18 attempts to reach Mars with unmanned probes over the years, the Russians have met with disaster 14 times.

In the 1950s and '60s, the Soviets were the true pioneers of space flight, and Mars was no exception to their ambitions. Beginning in 1960, they launched five Martian probes; two failed shortly after lift-off, two failed while in Earth orbit, and one lost contact with Earth while in interplanetary space more than 60 million miles away.

In 1962 at the height of the Cuban missile crisis, a nuclear war was nearly started by Sputnik 22, which had been intended to perform a Mars 'fly-by', but which broke up in Earth orbit instead. The debris falling back to Earth triggered US early warning systems, and brought the entire world to the edge of disaster.

By the 1970s, the Soviets were faring only slightly better: in 1971, the roving vehicle Mars 2 descended too steeply and crashed on the Martian surface, but its twin, Mars 3, managed to touchdown safely, only to cease transmitting seconds later. It is believed that the craft was overwhelmed by one of the massive dust storms which periodically envelop the planet's surface.

The Russians had experienced a similar disaster in March 1989, when their probe Fobos 2 was lost in Mars space. The Russian craft, however, managed to transmit some very intriguing – and unsettling – information before it was lost, information that might shed some light on what happened to the Mars Observer. While it was near the Martian moon Phobos, the probe photographed a large shadow on the surface. Moments later, it photographed the object that was apparently casting the shadow: a bright, cigar-shaped object that was calculated to have a length of 25 kilometres! The last images transmitted to Mission Control in Kaliningrad were of the object altering its attitude so that it was pointing directly at Fobos 2. Moments later, contact was permanently lost. If we take this information at face value, it seems at least possible that Fobos 2 was disabled or destroyed by the gigantic, anomalous object it had photographed, either because it had entered the region of Mars, or of Phobos itself.

In fact, Phobos is something of a mystery in its own right. First discovered, along with its companion Deimos, in 1877 by the astronomer Asaph Hall at the United States Naval Observatory in Washington, DC, the Martian satellite does not behave in quite the way a natural object should. Nearly 70 years later, in 1944, B.P. Sharpless, who also worked at the US Naval Observatory, attempted to determine the orbits of Phobos and Deimos. He discovered that the orbit of Phobos (the inner moon) was gradually decaying, a phenomenon also known as 'secular acceleration', whereby an object's velocity gradually increases until it enters the atmosphere. Natural bodies such as moons do not undergo secular acceleration; however, artificial satellites do.

The idea that Phobos might be an artificial satellite was first put forward by the Russian astrophysicist I.S. Shklovskii in 1960. After re-examining the work done by Sharpless, the Russian came to the conclusion that the density of Phobos is one-thousandth that of water. His reasoning was that, since Mars has no magnetic field that could influence the moon, its orbit must be decaying due to atmospheric drag from Mars. But the planet's atmosphere is too thin to slow down the orbital velocity of a moon; hence Phobos must have an unnaturally low density. Shklovskii realised that there was a serious problem with this conclusion: no natural solid object could possibly be a thousand times less dense than water. Therefore, he concluded, Phobos must be hollow; and, since this likewise could not have occurred naturally, the moon must be artificial, the product of a colossal engineering project.

However, when the first Viking probe went to Mars, it was able to approach Phobos and measure the moon's mass. Shklovskii's calculations were shown to be very inaccurate: the moon was far denser than he had stated, although its density was still unusually low.

'GLASS TUNNELS'

In recent years a great deal of controversy has arisen over certain other features that have been photographed on the surface of Mars. The same high-definition imaging technology that revealed the Face to be an entirely natural landform has also revealed what appear to be enormous glass tubes snaking across the Martian landscape, and disappearing underground at certain locations.

One such structure was photographed at 39.12°N 27.08°W by the Mars Global Surveyor probe on 11 August 1999. The area contains several deep valleys or channels which give the appearance of having been created by flowing water in the distant past. The strange, worm-like structures lie within these channels, and they appear to have a ribbed texture, as if composed of thick rings with a near-circular cross section, spaced approximately 60 metres apart, with thinner semi-translucent material between them.

Perhaps even more astonishingly, in some places the 'tubes' or 'tunnels' appear to be anchored to the sides of the valleys in which they lie. Looking at the Surveyor photographs, it is easy to let one's imagination run riot, postulating all kinds of explanations for the presence of these intriguing objects. Are they, as some have suggested, the remains of an ancient transit system? Were astronauts to visit Mars and find a way into them, would they discover derelict 'carriages'?

Or are the 'glass tunnels' actually irrigation channels, providing ironic proof of Percival Lowell's suggestion, made at the turn of the twentieth century, that an ancient Martian civilisation might once have attempted to irrigate their dying planet with water from the poles?

Certainly the landforms are very large: their width ranges from 100 to 200 metres, and they stretch above ground for several kilometres before plunging into deep canyons and apparently vanishing underground – images which are reminiscent of the giant sandworms in Frank Herbert's science fiction classic *Dune*.

However, speaking of 'dune', there is a more prosaic explanation for these features. They could, after all, just be ranks of sand dunes blown into strange shapes by the Martian winds. This is, in fact, the explanation put forward by NASA and the planetary science community, and this would seem to be a prime example of the usefulness of Occam's Razor, the principle which states that the simplest explanation for a phenomenon is usually the correct one.

Whatever their true origin, the 'glass tunnels' of Mars are strikingly beautiful features, and one hopes that one day they will be explored in their entirety . . . even those sections that appear to plunge into the Martian landscape.

STRANGE STRUCTURES ON EARTH'S MOON

While Mars and its moons hold many fascinating mysteries, we needn't travel so far from home to find bizarre cosmic puzzles hinting at alien activity.

The next time you look at the full Moon, concentrate on the very centre of the disc. You will be looking at the region known as Sinus Medii, or Central Bay. This is one of the many places on the Moon where strange things have been seen. Although you won't see it without a good telescope, there is a 10-kilometre-wide crater here called Ukert, at the centre of which lies a perfect equilateral triangle. No one knows exactly what this feature is, or whether its presence has anything to do with the fact that Ukert is at the 'sub-Earth point', the central point on the Moon's disc as seen from Earth, and thus the point on the Moon closest to Earth.

Just south-west of the Sinus Medii region stand two objects that are simply staggering in their implications, not to mention their physical size. In 1967 NASA launched the unmanned Lunar Orbiter III to conduct a photo-reconnaissance of the Moon and search for suitable landing sites for the upcoming Apollo landings. During the mission, the probe photographed the objects that have come to be known as 'the Shard' and 'the Cube'. The Shard is a near-vertical column one kilometre high, with a peculiar bulge in its mid-section; within this bulge is a geometric feature which gives the impression of being hexagonal in shape. There appears to be no plausible geological explanation for the presence of this eerie object.

Even more impressive – although less easy to see without the aid of computer enhancement – is the Cube, a glass-like structure about 1.5 kilometres wide, apparently composed of a large number of smaller sub-cubes, all suspended in a darker, meteor-eroded matrix. This matrix forms what seems to be a near-vertical tower 11 kilometres high, supporting the Cube. Computer-enhanced false-colour images of the original photographs show that the most intense light scattering is within the interior of the Cube, not its exterior as one would expect from a natural geological feature. This implies that the Cube and its tower are not geological features, but are constructed of semi-transparent, meteor-eroded glass.

In the spring of 1994, the Pentagon sent an unmanned

military probe, Clementine, to photo-reconnoitre the entire lunar surface. The craft carried state-of-the-art multi-spectral cameras, capable of analysing the composition and distribution of minerals. The results of that mission, which were allegedly leaked to Richard Hoagland and his colleagues, apparently show that they were correct in their interpretation of the original Lunar Orbiter images. The titanic structures on the Moon do not appear to correspond to the principles of lunar geology, and seem to have been badly eroded by millennia of meteoric bombardment. This implies that they were once part of far larger and more extensive structures, which probably included vast domes to protect the now-ruined complexes below.

The theory that domes once extended several kilometres above the lunar surface seems to be borne out by the presence of an object known as 'the Castle', which Hoagland describes as a 'geometric glittering glass object hanging more than nine miles above the surface of the Moon'. The fact that this object seems to be hanging from a thin filament implies that it forms a part of some vast and elaborate framework.

Hoagland claims that the true purpose of President Kennedy's decision to get to the Moon by the end of the 1960s was inspired by an urgent desire to get American astronauts to the derelict lunar cities before the Soviets. Did the Apollo astronauts succeed? Did Alan Bean of Apollo 12 really walk amid 'tiers of glasslike ruins', as Hoagland claims one photograph shows? Bean himself says not. When questioned by reporters from CNN and the Associated Press, the veteran astronaut replied: 'I wish we *had* seen something like what he's describing. It would have been the most wonderful discovery in the history of humankind – and I can't imagine anyone, in my wildest dreams, not wanting to share that.'

According to Hoagland, however, Alan Bean and his fellow astronauts were simply being loyal to NASA, which still doesn't want the potentially catastrophic truth to be revealed to the world.

As might be expected, Hoagland's discoveries have been ridiculed by officialdom. Experts in geology such as Paul Lowman of the Goddard Space Flight Center believe the lunar structures are no more than image-processing effects. However, Hoagland has his own circle of equally eminent experts who support his findings. It seems that, until another unmanned mission is sent to the Moon, and its findings released to the public, the debate will continue. Apparently, Richard Hoagland is seeking private funding for just such a mission. Let us hope that he succeeds.

In the meantime, it is worth noting that some very clever fake film footage has recently been produced, purportedly showing Neil Armstrong and Buzz Aldrin of Apollo 11 investigating a gigantic hangar-like building on the Moon, as well as what appears to be a crashed alien spacecraft partially buried beneath the lunar surface. The latter footage is allegedly from the top secret 'Apollo 20' mission, which was sent to the Moon to investigate the alien city and other artefacts. These delightful hoaxes can be found on YouTube, by typing 'Apollo 20', 'alien moon city', or variations thereof, in the search field.

Enjoy!

11

REGION OF THE LOST

THE BERMUDA TRIANGLE

Without doubt, the Bermuda Triangle is the greatest mystery of the ocean. The very name conjures images of aircraft flying into strange oblivion, their pilots frantically radioing base that something is terribly wrong, before losing contact for ever; of ships foundering in sudden, unnatural storms before plunging to an unimaginable fate on the hidden seabed. Countless books and articles have been written about it, and it has featured on countless TV shows dedicated to the enigmas of the paranormal.

There are those who say that the mystery of the Bermuda Triangle may never be solved; and there are others who say that it is no mystery at all, and that ships and planes have always met with tragic but quite natural disaster, both there and in many other places across the globe. Travel on the sea and in the air, they say, has always been dangerous, and probably always will be, and the Bermuda Triangle is no more sinister or perilous than any other area of ocean – despite what the febrile imaginations of writers, journalists and programme-makers would have us believe.

What is the truth about the Bermuda Triangle? Should we pay any heed to the strange and frightening concepts that have been associated with it over the years? Or should we, instead, believe the sceptics when they tell us that there is nothing 'paranormal' or 'supernatural' about the region – that, indeed, the 'paranormal' and 'supernatural' do not even exist?

A STRANGE CATALOGUE OF VANISHINGS

Perhaps we should begin with the words of the writer most closely associated with the Bermuda Triangle, Charles Berlitz, who did more than anyone else to bring its mysteries to a worldwide audience:

> Large and small boats have disappeared without leaving wreckage, as if they and their crews had been snatched into another dimension . . . in no other area have the unexplained disappearances been so numerous, so well recorded, so sudden, and attended by such unusual circumstances, some of which push the element of coincidence to the borders of impossibility.

The Triangle itself extends south-west from the island of Bermuda to Miami, Florida, then south-east through Puerto Rico, and finally northward back to Bermuda. The name 'Bermuda Triangle' was coined in 1964 by the writer Vincent Gaddis, but other writers have called it different (although slightly less elegant) names, such as the Devil's Triangle and the Limbo of the Lost. Indeed, they have altered not only its name, but also its extent, based on their own investigation of air and sea disasters. The largest of these additional triangles, the Limbo of the Lost, extends from Miami past Barbados to the north-east coast of Brazil, before turning through a near-right-angle and extending right across the Atlantic Ocean to the west coast of Ireland, and finally returning to Miami.

The researcher Paul Begg notes that Bermuda has always had an 'evil reputation'. Discovered by Juan de Bermúdez in 1515, the 300 or so tiny islands making up the group would seem to have been a perfect staging point for the long, dangerous journey from Europe to the New World, with their gentle climate and plentiful fresh food and water. Nevertheless, says Begg, they were still shunned for a century following their discovery. 'They were feared by the tough Elizabethan sailors, Shakespeare called them "the still-vex'd Bermoothes", and they gained an evil reputation as a place of devils. No one knows why.' Could it be, as Begg suggests, that even then they were known to be part of a region where ships and men were snatched out of the world by powerful, unknown forces?

We can choose the year 1800 as a convenient place to begin our regrettably but necessarily brief catalogue of modern losses in the Bermuda Triangle. In that year the USS *Pickering* disappeared with all hands during a voyage to the West Indies. Four years later the British ship *Bella* vanished while sailing from Rio de Janeiro to Jamaica (although Begg reminds us that she was dangerously overloaded, and may simply have capsized).

In August 1840, the merchant ship *Rosalie* sailed through the Sargasso Sea north-east of Haiti, bound for New Orleans. She was found undamaged but abandoned in the Bahamas. The fate of her crew has never been discovered. An account of the incident appeared in the *London Times*; the reporter (who mistakenly claimed that the ship was bound for Havana) stated:

> A singular fact has taken place within the last few days. A large French vessel, bound from Hamburg to Havannah, was met by one of our small coasters, and was discovered to be completely abandoned. The greater part of her sails were set, and she did not appear to have sustained any damage. The cargo, composed of wines, fruits, silks, etc., was of a very considerable value, and was in a most perfect condition. The captain's papers were all secure in their proper place . . . The only living beings found on board were a cat, some fowls, and several canaries half dead with hunger. The cabins of the officers and passengers were very elegantly furnished, and everything indicated that they had only recently been deserted. In one of them were found several articles belonging to a lady's toilette, together with a quantity of ladies' wearing apparel thrown hastily aside, but not a human being was to be found on board. The vessel, which must have been left within a very few hours, contained several bales of goods addressed to different merchants in Havannah. She is very large, recently built, and called the *Rosalie*. Of her crew no intelligence has been received.

The British training ship *Atalanta* went missing in 1880 with 290 cadets and crew, and four years later the Italian schooner *Miramon* also sailed into oblivion. In 1902, the German barque *Freya* suffered a similar disaster to the *Rosalie*. While en route from Manzanillo, Cuba, to Punta Arenas, Chile, she was discovered drifting in the Triangle, with her crew missing. Unlike the *Rosalie*, however, this ship was badly damaged, was listing badly and was partially dismasted, as if she had recently been through a severe storm. The strange thing was that there had been no recent storms in the area. If a storm was not to blame, what could have damaged the vessel so badly? And what happened to her crew?

A very slight, very tantalising clue is presented by the fate of the Japanese freighter *Raifuku Maru*, which went missing in 1925, but not before her radio operator managed to send this brief message: 'Danger like dagger now. Come quick!'

'What kind of danger looks like a dagger?' asks Paul Begg. 'Was dagger the only comparison the terrified radio operator could draw to the unworldly something that threatened and eventually took his ship?' Today, we might liken a missile or a sophisticated fighter jet to a dagger, but it is difficult to see what could have drawn such a comparison in 1925.

VANISHING AIRCRAFT

By far the most famous disappearance in the Bermuda Triangle is that of Flight 19. On the afternoon of 5 December 1945, five US Navy Grumman TBM Avenger torpedo-bombers took off from Fort Lauderdale, Florida, on a routine two-hour training flight over the Atlantic. In command was Flight Leader Charles Taylor; the other four pilots were trainees who were in the process of clocking up flight hours without instructors on board.

It was a beautiful day for flying, with near-perfect visibility. Their course would take them 123 miles east over the Bahamas, just an hour away at their cruising speed of 140 mph.

At 3.45 p.m. the control tower at Fort Lauderdale received a radio message from Taylor: 'This is an emergency. We seem to be off course. We cannot see land . . . repeat . . . we cannot see land.'

The tower asked him for their position.

'We're not sure of our position. We can't be sure where we are. We seem to be lost.'

The tower ordered them to head due west.

'We don't know which way is west. Everything is wrong . . . strange. We can't be sure of any direction. Even the ocean doesn't look as it should.'

At this point, another flight instructor, Lieutenant Robert Cox, who was flying in the vicinity of Fort Lauderdale, overheard the exchange, and radioed Flight 19, telling Taylor that he was going to change course, meet them and guide them back to base.

However, after a few moments of silence, it is claimed that Taylor said: 'Don't come after me. They look like . . .'

That was the last message received from Flight 19. The time was now 4.30 p.m. A Martin Mariner seaplane was despatched to look for the Avengers. It sent one radio message, and was never heard from again. In spite of a huge air–sea search, no wreckage of any of the planes or their crews was ever found.

The above account of the disappearance of Flight 19 is the version most often presented in books on the Bermuda Triangle. But how accurate is it? Perhaps we should look a little more carefully at the events of that tragic day.

Let's start with the weather. It was indeed bright and clear when the Avengers took off, but conditions gradually deteriorated during the flight. In addition, since this was a training flight, none of the pilots (with the exception of Taylor) was experienced: they each had only about 300 flying hours. For his part, Lieutenant Taylor, while an experienced pilot with more than 2,509 flight hours under his belt, had recently moved to Fort Lauderdale from Miami, and was unfamiliar with the area.

Then there is that strange radio exchange between Taylor and Lieutenant Cox. The implication to be found in most accounts of the disaster is that Taylor saw something terrifying in the sky, and warned Cox not to come after him. His final words, 'They look like ...', are particularly spine-tingling. However, the actual exchange between the two pilots went like this:

Cox: What is your trouble?
Taylor: Both my compasses are out and I am trying to find Fort Lauderdale, Florida. I am over land, but

it's broken. I'm sure I'm in the Keys, but I don't know how far down and I don't know how to get to Fort Lauderdale.

COX: Put the sun on your port wing if you are in the Keys and fly up the coast until you get to Miami, then Fort Lauderdale is 20 miles further, your first port after Miami. The air station is directly on your left from the port. What is your present altitude? I will fly south and meet you.

TAYLOR: I know where I'm at now. I'm at 2,300 feet. Don't come after me.

COX: Roger. I'm coming up to meet you anyhow.

The claims of some writers notwithstanding, Taylor never said 'They look like ...', and there is no indication from the radio exchange that anything out of the ordinary was occurring outside the planes.

However, Taylor was not able to reorientate himself. As Begg states: 'Many factors contributed to his disorientation: his compasses were not working, or he believed they weren't; he didn't have a clock or watch; his radio channel was subject to interference from Cuban radio stations, but the fear of losing contact with the flight deterred him from changing frequencies to the undisturbed emergency channel.'

Nor were the words 'don't come after me' the last to be transmitted. As time wore on, and the planes flew first one way and then another, it became clear to Taylor that they would soon run out of fuel, and would have to ditch in the sea. Night was drawing on as he radioed his men: 'All planes close up tight . . . we will have to ditch unless landfall . . . when the first plane drops to ten gallons we all go down together.'

And that, it seems, is precisely what happened. The Martin Mariner rescue plane did not fly 'into oblivion' – at least, not the type of oblivion most writers on the Bermuda Triangle have suggested. In fact, the seaplane is believed to have exploded shortly after take-off, an event which was witnessed by the crews of two ships, a freighter called the *Gaines Mills* and the USS *Solomons*. The cause of the explosion is unclear, but since the aircraft was carrying a large amount of fuel, and since fuel fumes were known to collect inside the fuselages of Mariners, it could easily have been caused by something as simple as an electrical fault.

The mystery of Flight 19 is not really a mystery at all, but a tragic accident whose main events are recorded and available to anyone who takes the trouble to do a little research. Most writers on the subject, however, prefer to maintain the myth of some kind of supernormal event having overtaken the unfortunate pilots. As so often happens in the literature of the paranormal, each writer who cites a particular case embellishes it just a little, and the next writer cites the previous writer's work as being a true and faithful account of what actually happened. After 10 or 20 or 30 writers have described the event, one can well imagine how far from actuality the accounts can drift, just as the poor pilots of Flight 19 drifted from their own true course.

A FINAL THOUGHT

In spite of the fact that many of the mysteries associated with the Bermuda Triangle have been solved, many people still believe that something truly abnormal is happening in the region. The theories read like the Golden Age science fiction of writers such as E.E. 'Doc' Smith and A.E. Van Vogt, and range from UFOs shooting down planes, to time warps and dimensional gateways, to still-functioning machines in sunken Atlantis causing havoc with vessels' electrical systems.

The reason for this is straightforward enough: bizarre theories and spine-chilling 'revelations' are *fun to read*. Many readers (and the present author confesses to being one of them) buy such books for the sheer entertainment value. It is a great pleasure to settle down with a book describing rampaging aliens, weird monsters, ravenous time portals and other such strangeness, and to wonder if just a little of it might be true, even while shaking one's head at the sheer foolishness of it all.

It is strongly to be suspected that many writers of such books realise this also, and are catering to a readership that is much more sophisticated than the sceptics might assume. It is likewise seriously to be doubted that their audience believe absolutely everything that they read in these books. Some do, of course (and they must live in a terrifying world indeed!), but the majority are surely mature and intelligent enough to sort fact from fantasy.

12

RIDDLES OF THE UNIVERSE

DARK MATTER AND DARK ENERGY

It seems that there's more to the Universe than meets the eye – quite literally.

Next time you get the chance, look up at the night sky (assuming, of course, that the urban light pollution where you live isn't so great that you can't even *see* the night sky). If you're lucky enough to get an unimpeded view of the heavens, take a few moments to look at the stars. There are many more out there than you can see with the naked eye; in fact, there are many more than can be seen with a powerful telescope. To say that the Universe is immense is to utter the understatement of all time.

The few scattered thousands (at most) of bright pinpoints you may see ranged across the dark celestial vault are but the tiniest fraction of what's really out there. It's like the difference between what's in your back garden, and what's in the Amazon rainforest. Our Galaxy, the Milky Way, is by no means unusual, and yet it is 100,000 light years in diameter. It contains at least 100,000 million stars – and some estimates put this number as high as 500,000 million. To get some idea of just how big these numbers are, think of an Olympic-size swimming pool, and then imagine it filled with sand. If each grain of sand were a star, that's how many stars there are in our Galaxy (give or take a few million).

And by a curious coincidence, that's roughly how many galaxies there are in the observable Universe. Imagine that sand-filled swimming pool, with each grain of sand now representing a galaxy (each containing, of course, hundreds of billions of stars). There's an awful lot of stuff out there. But the countless trillions of planets, stars, nebulae and galaxies do not comprise *everything* that there is in the Universe: unbelievable as it may sound, about 90 per cent of the matter in the Universe is unaccounted for. The things we can *see* account for only a fraction of what exists; the rest is, for all practical purposes, invisible. The rest is dark matter.

What is dark matter? It sounds like something one might read in a science fiction novel. And yet, it is a question that has perplexed astronomers since the 1930s, and seems set to continue to cause befuddlement well into the new century. It all has to do with gravity.

Gravity is what holds the majestic structures of star systems and galaxies together. Just as the Earth and the other planets of our (or any other) planetary system are held in their orbits by the pull of the Sun's gravity, so the millions upon millions of stars in galaxies throughout the Universe are held in *their* orbits by gravity, as they follow their aeon-long paths around the galactic hub, the huge bright bulge at the centre.

So far, so good. But the problem facing astronomers is that the galaxies they observe are rotating in a certain way. If you stir a cup of coffee and then pour in some cream, you will notice that the cream becomes stretched and spread out as it is carried around by the circular motion of the coffee. This is how galaxies should move, with their spiral arms gradually elongated by their rotation. The problem is, this is precisely how they *don't* move. Instead, the rotational speed of a spiral galaxy like our own Milky Way is constant, right across the vast disc. It's as if the billions of stars were painted on a rigid disc, spinning silently in the limitless void of space. There seems to be only one solution to this conundrum: the galaxies are actually embedded in a far larger 'halo' of invisible material, and are carried around in its gravitational grip. Astronomers have calculated that there is ten times as much matter in and around a typical spiral galaxy as can be directly detected with their instruments.

Dark matter also exists on the largest cosmic scales, influencing the movements of entire clusters and superclusters of galaxies – the largest structures in the known Universe. The galaxies within these gargantuan agglomerations appear to be moving too fast to be held in place merely by the gravity of the material of which they are composed. We are in the ridiculous position of being unable to detect (or even explain the nature of) about 90 per cent of our Universe!

So, what could dark matter be? There are at least two possibilities. The first is that it is made up of strange objects known as 'brown dwarfs'. Brown dwarfs are essentially failed stars: giant planets with around 20 times the mass of Jupiter, which get hot, but nowhere near hot enough (50 million °C) to trigger the process of nuclear fusion that powers the stars and makes them shine. Eventually, brown dwarfs cool and fade completely, becoming black dwarfs – which are even harder to detect.

The second possibility involves very exotic particles called WIMPS (weakly interacting massive particles). WIMPS possess mass, and thus interact through gravity with ordinary baryonic matter (a term that refers to everyday atomic matter, which is composed of protons, neutrons and electrons), but otherwise interact very weakly with it. Such particles could account for the way the galaxies formed in the early stages of the Universe, clumping together to form regions of higher gravity around which baryonic material collected.

There is currently one known particle which could be a component of dark matter: the neutrino. Although the mass of a neutrino is vanishingly small (about one fifty-millionth the mass of a proton), neutrinos outnumber baryonic particles by about a billion to one. They exist in the Universe in such huge numbers that they might well account for the observed effects of dark matter. Unfortunately, they are also incredibly difficult to detect. A single neutrino could pass through a piece of lead the size of a galaxy without interacting with any of its atoms.

It seems that we will have to wait a while longer before we can pin down the identity of dark matter, and finally figure out why so much of our Universe remains hidden from us.

If dark matter sounds like something from science fiction, dark energy has all the makings of a surrealist fantasy. Ever since the late 1920s, when Edwin Hubble examined the redshifted light from distant galaxies and realised that the Universe was expanding, astronomers and cosmologists have maintained that the Universe began approximately 14 billion years ago in an unimaginably powerful fireball – the Big Bang – and its expansion has been gradually slowing down ever since. While the Big Bang is widely accepted as the mechanism by which the Universe – and we – came into being, the idea that its expansion is being gradually slowed down by the gravitational attraction of the matter which it contains has recently taken a serious knock.

Far from slowing down, the expansion of the Universe actually seems to be accelerating, due to a bizarre and ill-understood phenomenon called dark energy. The puzzle first became apparent in 1998, when two groups of astronomers conducted a survey of supernovae (exploding stars) in a number of distant galaxies. The supernovae were much dimmer than expected, which meant that their galaxies were much more distant than expected. The only explanation was that the expansion of the Universe had *sped up* at some time in the past. The conclusion was as shocking as it was inescapable: far out in the depths of intergalactic space, some mysterious force was acting against the pull of gravity, causing the galaxies to fly away from each other at ever-greater speeds.

Of course, being rightly cautious, the astronomy community assumed that there must be a simple explanation. Perhaps the supernovae were dimmer than they should have been because their light was being blocked by clouds of interstellar dust (there's a lot of it out there). Or perhaps the supernovae themselves were intrinsically dimmer than the astronomers had believed. But with careful checking and the gathering of more data, those rational explanations have been largely abandoned, leaving the irrational explanation: that a form of energy exists in the Universe which counteracts the effects of gravity. (As Sherlock Holmes once remarked, when you have eliminated the impossible, whatever remains, however improbable, must be the truth.)

The dark energy hypothesis is an echo of a theory formulated by Albert Einstein, which he called the 'cosmological constant'. Before Hubble's observations demonstrated that the Universe is expanding, Einstein found himself wondering why, if the Universe was static and unchanging, it didn't collapse in on itself under the pull of its own gravity. He suggested that it was held in balance by a force acting against gravity. When the expansion of the Universe was discovered, Einstein gratefully laid aside the cosmological constant, calling it the 'biggest blunder' of his career.

Astronomer Virginia Trimble, of the University of Southern California at Irvine, has a fairly simple way of imagining the phenomenon of dark energy:

> If you think in terms of the Universe as a very large balloon, when the balloon expands, that makes the local density of the [dark energy] smaller, and so the balloon expands some more . . . because it exerts negative pressure. While it's inside the balloon it's trying to pull the balloon back together again, and the lower the density of it there is, the less it can pull back, and the more it expands. This is what happens in the expanding Universe.

The supernova evidence suggests that this acceleration first manifested about five billion years ago. At that time, the galaxies were far enough apart that their gravity was overridden by the repulsive force of dark energy. Since then, the constant push of the dark energy has been causing the expansion to speed up; and it looks like the expansion will continue indefinitely.

Combined with the mysteries of dark matter, dark energy places astronomers and cosmologists in a rather invidious position: after all, they have devoted their lives to understanding the Universe and the processes that make it work. Now they have *two* bizarre phenomena which have a profound effect on the origin, nature and evolution of the Universe. Says astronomer Richard Ellis of Caltech: 'I'm as big a fan of dark matter and dark energy as anybody else, [but] I find it very worrying that you have a Universe where there are three constituents, of which only one [ordinary matter] is really physically understood . . . This isn't really progress.'

There's no doubt that dark energy has made life a whole lot more complicated for cosmologists. According to Craig Hogan, an astronomer at the University of Washington in Seattle: 'It's important to realise that dark energy is different from any other kind of energy we've ever found. Presumably, if we ever get a truly unified theory of everything, which includes gravity and the other forces of nature, one of the big tests of that is, does it predict dark energy? Does it get that right or not?'

Richard Ellis stresses the importance of confirming the supernova results beyond any doubt.

> If the supernova results were not to hold up . . . because supernovae were found to be different at earlier times – perhaps they were dimmer for some reason we don't understand – then the Universe wouldn't be accelerating. I think it's unlikely, but I think it's so important that we have to check. It's such a big claim, and it's so counter-intuitive that the Universe would be accelerating, in my opinion, that no stone should be left unturned. We should verify this as best as we possibly can.

13

Messages from the Dead

Electronic Voice Phenomena

In 1901 the American ethnologist Waldemar Bogoras travelled to Siberia to study the Tchouktchi tribe. During his visit, Bogoras was invited to observe a spirit-conjuring ritual, in which a shaman beat upon a drum with increasing rapidity until he entered a trance state. Presently, Bogoras was surprised to hear strange voices filling the room. The voices seemed to come from everywhere around him and, even more startlingly, some of them spoke in English.

After the shaman had completed the ritual, Bogoras asked if he could be present at the next spirit conjuring, and the shaman agreed. The ethnologist later wrote:

> I set up my equipment so I could record without light. The shaman sat in the furthest corner of the room, approximately twenty feet away from me. When the light was extinguished the spirits appeared after some 'hesitation' and, following the wishes of the shaman, spoke into the horn of the phonograph.

When Bogoras played back the recording, he could distinguish a clear difference between the speech of the shaman, audible in the background, and the spirit voices which seemed to have been speaking directly into the mouth of the phonograph's horn. Throughout the ritual, the shaman's constant drumbeats could be heard, issuing from the same spot in the room.

Thus concluded the first-known experiment in which the voices of 'spirits' were captured on a recording device.

In the 1920s, the great inventor Thomas Alva Edison, who held more than 1,300 US and foreign patents for various electrical devices, attempted to design a machine which would allow communication with the spirits of the dead. Edison's assistant, Dr Miller Hutchinson, wrote: 'Edison and I are convinced that in the fields of psychic research will yet be discovered facts that will prove of greater significance to the thinking of the human race than all the inventions we have ever made in the field of electricity.'

Edison himself wrote:

> If our personality survives, then it is strictly logical or scientific to assume that it retains memory, intellect, other faculties, and knowledge that we acquire on this Earth. Therefore . . . if we can evolve an instrument so delicate as to be affected by our personality as it survives in the next life, such an instrument, when made available, ought to record something.

Unfortunately, Edison died before he could develop his communication device. As he lay on his deathbed, he remarked to his physician: 'It is very beautiful over there . . .'

THE PIONEERS

In 1959 the Swedish singer and film producer Friedrich Jürgensen was out walking one day. It was his habit to take a tape recorder with him on such occasions, for he loved the sound of birdsong, and he frequently made recordings. When he played back one of the tapes, Jürgensen was surprised to hear more than the musical chirpings of his beloved birds: he also heard a voice speaking in Norwegian. Naturally, his first assumption was that the tape recorder was acting as a radio receiver, and had picked up the transmissions from a nearby station. He thought no more about the curious – but apparently explainable – phenomenon.

A few weeks later, however, he was playing back another birdsong recording, when he heard a woman's voice saying in German: 'Friedel, my little Friedel, can you hear me?' He immediately recognised the voice as that of his mother, who had died several years earlier.

Jürgensen began to experiment with other ways to receive these apparent voices of the dead, and came up with the idea of detuning an ordinary radio receiver so that it received nothing but 'white noise'. He wrote up the results of these experiments in 1964 in a book entitled *Rösterma Från Rymden (Voices From the Universe)*.

Jürgensen's book came to the attention of Dr Konstantin Raudive, a Latvian psychologist, who visited Jürgensen the following year. Raudive (after whom the phenomenon would later be named 'Raudive voices') conducted his own experiments and, after months of silence, at last picked up a message in Latvian. Raudive would go on to receive more than 70,000 such messages over the next three years. In 1968 he published his findings in a book entitled *Unhörberes wird Hörbar (The Inaudible Made Audible)*, which was translated into English in 1971 as *Breakthrough*.

In 1972 sound and radio engineers across Europe declared their interest in what had come to be known as Electronic Voice Phenomena (EVP). The chief engineer at Pye offered the opinion that the phenomena were worthy of serious investigation. R.K. Sheargold, then Chairman of the Society for Psychical Research's Survival Joint Research Committee (SJRC), investigated the phenomena and succeeded in capturing voices. Sheargold decided to write to Colin Smythe Ltd, the publishers of *Breakthrough*, confirming that he was in a position to assure his colleagues in the SJRC that the phenomena were indeed genuine.

Between 1970 and '72, the Society for Psychical Research commissioned one of their members, D.J. Ellis, to investigate EVP. His conclusions, however, were less than enthusiastic: in his opinion, the phenomena were natural rather than supernatural, and were the result of random noises being interpreted as voices. Indeed, it must be admitted that EVP voices are not always particularly clear, and those listening to them often hear different 'words' being 'spoken'.

However (and this echoes Charles Fort's dictum that for every expert there is an equal and opposite expert), other researchers claimed to have gone through the messages syllable by syllable (in isolation from each other), and to have discovered extremely close correspondences in their interpretations.

There are two main problems with Electronic Voice Phenomena. The first is that electronic equipment of any kind can pick up radio signals, and no amount of screening can prevent this possibility from undermining – potentially at least – all such recordings. The second difficulty is that, in the case of 'white noise' recordings, the messages are almost drowned out by the surrounding static, making it extremely difficult to interpret them accurately. Sceptics maintain that what is being picked up in these cases is no more than the occasional electronic glitch. Supporters of EVP counter that the messages received are often relevant to the listener, which rules out both random radio reception and electronic glitches.

Perhaps the most interesting aspect of EVP is that experiments can be done by anyone: no equipment more sophisticated than a simple radio receiver and a tape recorder is required (at least to begin with), and the phenomena offer a potentially fascinating way for the average person to do a little psychic research for themselves.

For those wishing to capture Electronic Voice Phenomena, a good-quality tape recorder is essential. In previous years, researchers used reel-to-reel recorders; but these days such equipment is quite rare, and a good cassette recorder is more than adequate for the job. When buying a unit one should, however, ensure that it has the following features: a forward/backward key; a rewind key (for quick repetitions without stopping the tape first); and a speed regulator. This last feature is especially important to an accurate interpretation of the results, since occasionally the voices captured speak either very quickly or very slowly.

The methodology of attempting to capture EVP is quite straightforward. When one is ready to begin the experiment, all one has to do is hit the record key and introduce oneself, speaking clearly into the microphone. One can either address a certain deceased individual directly, or those in the 'spirit realm' generally. It is, of course, best to leave pauses between questions, to allow time for answers to be formulated. Once the session has ended, the tape should be rewound to the beginning, and then played, with the experimenter listening carefully for voices.

The alternative method is to utilise the white noise of a detuned radio, connected via cable to the cassette recorder. In her book *Bridge Between the Terrestrial and the Beyond*, the EVP researcher Hildegard Schaefer explains why detuned radios can be very useful in capturing the voices:

> The reason for giving preference to the recording with the aid of a broadcast receiver is that a multitude of energy, pulsations and frequencies is made available to the partners on the Beyond so that recordings of better quality and quantity can be achieved. Experience teaches that with this method the voices manifest more frequently, are in fact louder and thus better understandable.

We are constantly reminded that we live in an 'information age'; and there is certainly no doubt that computer technology has changed our world in vastly significant and irreversible ways. We are utterly dependent on computers, which now regulate or control virtually every aspect of our lives. Assuming that the dead occasionally try to make contact with the living, it would seem quite reasonable for them to attempt to do so using any means available here in the material world, taking advantage of the latest developments in information technology.

CONTACT VIA COMPUTER?

The *Swiss Bulletin for Parapsychology* for November 1986 included an article on the strange events that occurred at the home of a man named Ken Webster near Chester, England. According to the article, a discarnate personality from the sixteenth century had made contact with Webster the previous year, via Webster's computer. Webster, a college teacher, had bought and renovated the house in 1984, whereupon strange and inexplicable manifestations began to occur, suggesting the opening phases of a classic haunting. Furniture was moved around, various household objects disappeared, and other psychokinetic phenomena alerted Webster that something very unusual was going on.

The supernatural activity was not limited to objects in the house, however. Messages began to appear on Webster's computer, both on the screen and on diskettes. Eventually, the number of messages grew to more than 250.

At a loss to explain what was happening, Webster showed some of the mysterious messages to a colleague, an English teacher named Peter Trinder. After a lengthy examination, Trinder concluded that they were written in English of the fourteenth to sixteenth centuries. The messages were apparently written by a man named Thomas Harden, who claimed to have lived at the time of Henry VIII. Harden claimed to have been educated at Brasenose College, Oxford, a claim that Webster and Trinder later verified.

Webster occasionally tried to catch Harden out. For instance, he tapped the question 'Is King James on the throne?' into his computer (which Harden called a 'light box'). The reply was: 'The King is of course Henry VIII. I don't know of any King James.'

The Webster–Harden computer correspondence was investigated by the parapsychologist Dr Theo Locher, who concluded in the *Swiss Bulletin for Parapsychology*:

> It seems possible that an unredeemed soul, still bound to the place of its distress, who has not realised its passing into the Beyond, has the ability to suitably operate a computer by its thoughts and ideas psychokinetically. Under certain conditions, reciprocal effects are possible.

An alternative possibility is that Harden was not actually dead at all, and that he was somehow able to communicate with Webster while still alive in his own time. This would be an example of a phenomenon known as a 'time slip' (see Chapter 26), whereby a channel is spontaneously opened between two periods of history.

SPIRICOM

Sceptics are forever complaining about the lack of conclusive evidence for the paranormal, evidence that can be repeated under controlled laboratory conditions by scientists on opposite sides of the world. It is ironic that for more than two decades there has been just such a body of evidence in the form of a refinement of EVP research known as 'Spiricom'.

In August 1981, the investigative journalist John G. Fuller (who had already achieved international fame with his groundbreaking investigations into the alleged alien abduction of Barney and Betty Hill, and the so-called 'ghost of Flight 401' – see Chapters 24 and 15, respectively) received a curious letter from a man named George Meek, who was head of a foundation called Metascience, based in Franklin, North Carolina. A highly successful engineer and holder of numerous patents, Meek claimed to have headed a research effort to establish two-way communication with the discarnate personalities of humans who had died. The name of this research effort was Spiricom.

Against his better judgement, Fuller became more and more interested in this apparently outrageous claim, and eventually wrote a book on the subject, *The Ghost of 29 Megacycles*, which was published in 1985.

Meek's research proceeded along a strange path that included elements of spiritualism and engineering. In the early days, he had become acquainted with a man named Bill O'Neil who, while lacking a conventional education, possessed an innate genius for electronics. In addition, O'Neil was an extremely sensitive psychic and healer. The two men combined their talents to enlist the aid of deceased scientists in the design and construction of an electronic device which would enable two-way communication with the afterlife.

O'Neil lived with his wife on an isolated farm in western Pennsylvania, where he conducted his psychic and scientific experiments. These experiments resulted in contact with a discarnate person calling himself 'Doc Nick', a deceased radio ham who appeared one day in O'Neil's cluttered laboratory. The spirit of Doc Nick initially appeared only as a disembodied head, shoulder and right arm. Understandably gripped with terror, O'Neil gasped: 'My God, who are you?'

The apparition responded by giving its name; it added that it had been a radio ham, and asked O'Neil for his own call-sign letters. The apparition concluded its first visitation by saying: 'I'm going to guide you along on a regular basis. Detailed suggestions. On a regular basis.'

O'Neil then began to receive help from another discarnate personality calling himself 'Dr Mueller', an altogether more irascible character, who seemed unaware of the existence of Doc Nick. Intriguingly, during one of his appearances in O'Neil's laboratory, Dr Mueller provided a great deal of information about his former life on Earth:

> Name: Dr George J. Mueller.
> Former Social Security number: 142-20-4640.
> Ancestry: English, Irish, German. BS in Electrical Engineering, University of Wisconsin. Top fifth of his class in 1928. MS in Physics, Cornell, 1930. PhD in Experimental Physics at Cornell, 1933. Additional training, New York University and UCLA. Meritorious Civilian Award from the Secretary of the Army. Physics Instructor and Research Fellow while at Cornell.

O'Neil passed on this information to George Meek who, after some months of research, was able to verify every item on Dr Mueller's '*curriculum post mortem*'.

With the help of O'Neil, Doc Nick and Dr Mueller, George Meek and his Metascience foundation were ultimately able to construct a sophisticated radio transceiver, which was apparently capable of opening a channel of communication between this world and the afterlife, and of recording the results.

On 6 April 1982, George Meek held a press conference at the National Press Club in Washington, DC. In spite of his preference for total anonymity, Bill O'Neil also attended, as did John G. Fuller and about 30 reporters and journalists, including representatives from the Associated Press, United Press International, Reuters, *Business Week*, *Harper's*, National Public Radio and the Chicago *Sun-Times*.

After playing some sample tapes of the recorded conversations between the living and the 'dead', Meek announced that he would not be patenting the Spiricom device, and that plans would be made available to the public for a nominal fee to cover printing costs. The aim of the press conference, said Meek, was to bring to the attention of scientists around the world the potential that Metascience and Bill O'Neil had discovered for a colossal widening of humanity's awareness of the universe, and of its own destiny.

George Meek's work has been continued by researchers all over the world, most notably the American Association–Electronic Voice

Phenomena, Inc., which holds conferences in numerous countries. The association's mission is to 'Provide Objective Evidence That We Survive Death in an Individual Conscious State'.

A DIRECT LINE TO THE AFTERLIFE?

One of the most unsettling phenomena associated with EVP are the so-called 'phone calls from the dead', in which the spirit of a deceased person takes it upon him- or herself to initiate electronic contact with the living. In most cases, people receive calls from deceased loved ones, often on dates of special significance to the person.

There are certain characteristics that are often present in these strange telephone calls. The telephone usually rings normally, but sometimes it may sound strangely flat, as if heard from a great distance. The connection is usually quite bad (as one might expect!), with the voice on the other end of the line fading in and out; however, the voice is recognisable, and the speaker often makes statements or imparts information of a personal nature which no one else could have known. The call is often terminated abruptly, either by the caller hanging up, or the line going dead.

If the recipient knows that the caller is dead, he or she may well be too shocked to speak, and the caller will quickly hang up. In more intriguing cases, the recipient does not yet know that the caller is dead, and a conversation will ensue, sometimes lasting for as long as half an hour. Most phone calls from the dead occur within 24 hours of the caller's death, although some have been reported as long as two years after the time of death.

As to their purpose, it seems that the caller wishes to leave either a farewell message or a warning of impending danger. Occasionally, some piece of information needed by the living is given, as in the famous case of the actress Ida Lupino, who received a phone call from her father six months after his death, telling her the whereabouts of some papers needed to settle his estate.

A particularly poignant example of a phone call from the dead was originally reported by Don Owens of Toledo, Ohio, and appeared in the September 1969 issue of the US magazine *Fate*.

> Don had a very close friend, Leigh Epps, a bachelor who had no luck with women. Consequently, he valued his few close friends, who included Don and his wife,

whom he called 'Sis.' After Leigh moved to another area they drifted apart, and their meetings became rare. At 10.30 p.m. on 26 October 1968, while Don was out, his wife received a phone call from Leigh:

'Sis, tell Don I'm feeling real bad. Never felt this way before. Tell him to get in touch with me the minute he comes in. It's important, Sis.'

Don tried to phone Leigh back many times but with no success. He discovered later that at exactly the time of the phone call, Leigh had died in the Mercy Hospital, only six blocks away.

Equally bizarre and unsettling are the cases in which a living person calls a friend or relative, unaware that they are dead, and receives an answer! As might be expected, there is no satisfactory explanation for why telephone calls can be placed to and from the dead. It is possible that they occur by supernatural manipulation of the telephone mechanisms, while another theory holds that they are the result of unconscious psychokinesis on the part of the recipient.

VISUAL EVIDENCE

Visual images of ghosts are, of course, as old as photography itself; but more intriguing are the instances of visual contact with the 'Beyond', using video recording equipment. In 1984 a couple from Luxembourg, Maggy and Jules Harsch-Fischbach, began experimenting with EVP, and received messages from spirits on an electronic link nicknamed the 'Eurosignal Bridge'.

Two years after their experiments began, a high-pitched, computer-like voice came through on their radios, and asked them to disconnect their television aerial and switch the TV to an untuned channel. The Harsch-Fischbachs did as the voice asked, and then set up a video camera in front of the television.

Presently, the couple saw several images flash across the screen, moving too quickly for any detail to be discerned. After allowing the video camera to record for another ten minutes, they played back the tape at a reduced speed. Almost immediately, the image of a deceased relative appeared. He informed the Harsch-Fischbachs that a research team was being assembled in the afterlife, and suggested that they do the same on Earth.

The couple lost no time in founding the *Cercle d'Etudes sur la Transcommunication* (CETL), the Study Circle in Transcommunication. Almost immediately, a being calling itself 'Technician' made contact, telling them that it had never been human, never animal, and had never occupied a physical body:

> I am not energy and I am not a light being. You are familiar with the picture of two children walking across a bridge, and behind them is a being who protects them. That's what I am to you, but without the wings. You can call me Technician, since that is my role in opening up this communication bridge. I am assigned to Planet Earth.

The Harsch-Fischbachs and their colleagues in CETL developed an ongoing line of communication with Technician. In 1988, the entity informed them that their group had been selected by the spirits on the Other Side to help them improve the communication link between the worlds of matter and spirit. The entities called the effort Project Timestream.

As is often the case with spirit communications, the discarnate personalities who began to communicate with CETL had been famous in life. They included the nineteenth-century British explorer Sir Richard Burton, Thomas Edison (who had finally found a way of communicating between worlds – albeit from the Other Side) and the writer Jules Verne, along with well-known figures from the EVP community such as Friedrich Jürgensen and Konstantin Raudive.

Over the years that followed, the Harsch-Fischbachs' small apartment was visited by scientists and reporters from all over the world, who watched in fascination as the images continued to flit across the television screen. On one such occasion, the nineteenth-century chemist Henri Ste Claire de Ville told American and German researchers: 'It is our job as well as your job to set fire to minds – to set fire to minds in your world, and in that moment to try to master time.'

The Timestream operators on the Other Side transmitted many pictures of themselves at work. However, these images cast doubt on the veracity of the earthly researchers. For instance, when a spirit communicator named Swejen Salter took over from Technician in 1992, she transmitted an image of herself to a CETL computer screen. Unfortunately, the image of Swejen Salter looked uncannily like Maggie Harsch-Fischbach.

Another picture which CETL claimed had been transmitted from the Other Side appeared to show the film director George Cukor and Thomas Edison in the Timestream sending station (which looked very much like an ordinary Earth laboratory). The image of Edison, apparently sporting contemporary clothes, bore a striking resemblance to a photograph of the inventor taken just before his death in 1931.

Paranormal researchers Jenny Randles and Peter Hough have made the intriguing suggestion that the CETL researchers may be the victims of clever hoaxers who have tapped into their equipment and left subtle clues pointing to the origin of their hoax. For instance, it is claimed by the entities comprising Project Timestream that on a planet on the Other Side named Marduk (a god of ancient Babylon, and also the mysterious rogue planet named by maverick archaeologist Zecharia Sitchin in his book *The Twelfth Planet*), there is a vast river called the River of Eternity. According to CETL, Sir Richard Burton has often spoken of his adventures along the River of Eternity on Marduk.

The problem is that this is essentially the plot of Philip José Farmer's classic science fiction novel of 1971, *To Your Scattered Bodies Go* (the first volume of the famous Riverworld Saga), in which the main protagonist (Sir Richard Burton) dies and finds himself resurrected on the banks of a vast waterway snaking across the surface of an alien planet, the Riverworld.

As so often happens in the fields of the paranormal, fact seems to shade into fantasy, with no clear line of demarcation between the two. It is interesting that CETL researchers have not sought to make money from the material they have collected, and they apparently dislike publicity, preferring to let their research speak for itself. However, other researchers are still suspicious of them; and it seems that, in spite of the wealth of material that has been amassed by CETL and other researchers around the world, we are still no closer to an unequivocal answer to the ultimate question: what happens to us when we die?

14

IS THERE ANYONE OUT THERE?

MESSAGES FROM THE UNIVERSE

At half past three on the afternoon of 4 September 1953, a Mr C.W. Bradley was watching television in his London home. Suddenly and without warning, the programme he was watching was replaced with the signal of the American TV station KLEE-TV. During the same month, the same letters appeared on the monitor screens of a company called Atlantic Electronics Ltd in Lancaster. Of course, television images are transmitted across the Atlantic and around the globe every second of every day; however, what Mr Bradley and the scientists at Atlantic Electronics saw on their screens in 1953 was truly astonishing, for two reasons.

First, the programme signal from KLEE-TV had been transmitted three years *previously* from Houston, Texas. It was replaced with the signal KPRC-TV in July 1950, and the letters KLEE-TV were never again transmitted by any station on Earth. Second, television transmissions between the United States and Europe were impossible at that time, and would remain so until 1962 and the launch of Telstar, the world's first communications satellite.

This bizarre incident has never been satisfactorily explained, although attempts at a solution have included the theory that clouds of ionised gas might somehow have stored the signal for three years, and then released it back to the Earth's surface.

There is, however, another theory which has been seriously considered by astronomers: that the original signal from KLEE-TV was received by an extraterrestrial probe somewhere in deep space, and then retransmitted back to Earth. The reason? To let the people of Earth know that 'someone' out there is now aware of our existence, and is attempting to let us know that we are not alone.

Such an event would provide the answer to one of the most profound questions in humanity's history: are we alone in the Universe, or is the Earth just one of countless worlds on which life has arisen and thrived? The great religions can offer us little beyond doctrinal pronouncements that must be taken on faith, with no corroborative evidence one way or the other. For most of its history, Christianity has taught that our world is unique, and that the cosmos was created for us alone. Hinduism, on the other hand, takes the opposite view, seeing nothing blasphemous or heretical in the notion of a plurality of inhabited worlds scattered throughout an infinite Universe.

While the idea of detecting and communicating with extraterrestrial civilisations has been one of the key tropes of science fiction for a century or more, it is now considered a real possibility by many serious scientists and astronomers. Contact with alien beings elsewhere in our Galaxy would, without doubt, be the most significant event in human history; the ramifications for our culture, our philosophy, religion and science, are scarcely conceivable, and yet those who are conducting this great search tell us that the day of First Contact must eventually come . . . if we are not alone.

If . . .

In 1992 the United States National Academy of Sciences published a report on the coming 'Decade of Astronomy and Astrophysics' by a committee of 15 leaders in those fields. Of the search for extraterrestrial intelligence, the committee said, in part: 'Ours is the first generation that can realistically hope to detect signals from another civilisation in the Galaxy.' The effort 'is endorsed by the committee as a scientific enterprise. Indeed, the discovery in the last decade of planetary disks, and the continuing discovery of highly complex organic molecules in the interstellar medium, lend even greater scientific support to this enterprise.'

After giving intermittent financial support to SETI (Search for Extraterrestrial Intelligence), NASA agreed to assign the telescope at Arecibo in Puerto Rico (at 1,000 feet in diameter, the largest antenna in the world) to scan the most promising stars for radio signals. At the same time, NASA's global network of deep space antennae would survey the entire sky. The space agency decided to begin this epoch-making search on 12 October 1992, exactly 500 years after Columbus reached the New World.

THE MYSTERY OF THE LONG DELAYED ECHOES

If there are other civilisations out there in the depths of galactic space, other beings who are intelligent enough to broadcast radio transmissions which we might be able to detect, might they not also be capable of sending unmanned probes into the void as part of their own search for extraterrestrial intelligence? Our best estimates place the age of the Universe at approximately 15 billion years; that's a lot of time for life and intelligence to evolve, some of which might well be much further along the road than we are. It's certainly *possible* that one or more alien civilisations have sent probes into the dark immensity beyond their own planetary shores, and that one of those probes might have happened upon our Solar System with its single intelligence-bearing planet – Earth. Possible, yes . . . but how *likely* is it?

In the science journal *Nature* of 3 November 1928, there appeared a letter to the Norwegian physicist Carl Störmer from a radio engineer named Jorgen Hals. In part, the letter read:

> At the end of the summer of 1927 I repeatedly heard signals from the Dutch short-wave transmitting station PCJJ at Eindhoven. At the same time as I heard these, I also heard echoes. I heard the usual echo which goes round the Earth with an interval of about one seventh second as well as a weaker echo about three seconds after the principal echo had gone. When the principal signature was especially strong, I suppose the amplitude for the last echo three seconds later, lay between one tenth and one twentieth of the principal signal in strength. From where this echo

comes I cannot say for the present, I can only confirm
that I heard it.

Hals had accidentally discovered a phenomenon that would
later be called 'long delayed echoes' (LDEs). In the following
decades, scientists around the world studied the phenomenon,
but were unable to offer a satisfactory explanation for it. The
LDE phenomenon occurs when certain radio messages are
transmitted into space, and are then reflected back to Earth, the
two main 'reflectors' being the Earth's ionosphere and the Moon.
The intriguing thing is that some echoes are received on Earth
several seconds after transmission. Since radio waves travel at the
speed of light (approximately 186,000 miles per second), and the
distance between the Earth and the Moon varies from 221,460
miles to 252,760 miles, the longest time they should take to be
reflected back is 2.7 seconds. Any delay longer than that implies
the presence of a 'reflector' further out in space than the Moon.

In September 1928, Carl Störmer, to whom Hals had
addressed his letter, attempted to validate the radio engineer's
findings. He transmitted radio signals of varying lengths at
30-second intervals, and detected echoes from three to fifteen
seconds. Now, as noted, it would have taken 2.7 seconds for
the signals to be reflected back from the Moon, and more than
four minutes to be reflected from Venus, the Earth's nearest
planetary neighbour. The clear implication was that the signals
were being reflected by something between the Moon and the
nearest planet.

Scientists have proposed several possible natural explanations
for LDEs, such as the hypothesis that the echoes are bouncing
back to Earth from streams of solar plasma. The Sun is constantly
radiating charged particles, and when these particles shed
electrons, the result is plasma (examples seen on Earth include
lightning and the spectacularly beautiful Aurora Borealis, or
Northern Lights). It is well known that charged particles in
plasma reflect radio signals, and since the Sun is throwing off a
vast amount of plasma in the form of the so-called 'solar wind', it
is at least conceivable that the LDEs are merely reflections from
these plasma streams.

In 1974 Anthony Lawton of the British Interplanetary Society
proposed that LDEs might result from radio signals being

trapped in the Earth's ionosphere and then reflected back to the surface. However, as critics of the hypothesis have pointed out, this might conceivably explain LDEs of a few seconds, but it stretches the limits of plausibility with echoes received after several minutes or longer.

Störmer was even more intrigued by the different time delays in the reception of his signals. The delay variations after one set of transmissions read, in seconds: 8–11–15–3–13–8–8–12–15–13–8–8. There seemed to be only two possible explanations for this: either there were a number of reflective objects out in space at varying distances from Earth, or there was a single object re-transmitting the signals at different time intervals.

But what could this mysterious 'transmitter' be?

A possible answer was put forward by the renowned radio astronomer Professor Ronald Bracewell, in an article which appeared in *Nature* in 1960. Bracewell noted that for most of its four-and-a-half-billion-year history the Earth would have been utterly uninteresting to any extraterrestrial intelligence searching for a civilisation with which to communicate. He suggested that superior civilisations would send automated messenger probes to stars with potentially life-bearing planets, to await the possible development of intelligent beings.

Each of these probes would draw its power from the star to which it had been assigned, and would be incredibly sophisticated, containing a computer so powerful it would possess many of the characteristics of an intelligent being. In a lecture he gave in 1962 at the University of Sydney, Bracewell stated: 'If we contemplate the resources of biological engineering, which we have not begun to tap yet, it is conceivable that some remote community could breed a subrace of space messengers, brains without bodies or limbs, storing the traditions of their society, mostly to be expended fruitlessly but some destined to be the instruments of the spread of intragalactic culture.'

While it is certainly entertaining to speculate on the future of spaceflight, as popularly represented by the fabulous interstellar vehicles of *Star Wars* and *Star Trek*, in dealing with the science of SETI we are constrained by what we know right now of the laws of physics. And what we now know implies that faster-than-light travel is either impossible or extremely difficult. (As the astronomer and founder of SETI, Frank Drake, says:

'Interstellar travel may turn out to be so difficult, no intelligent species would be dumb enough to bother.') The question, then, is: would an intelligent species consider it worth its while to send large, passenger-carrying vehicles to explore other star systems?

The answer may well be: no.

Such flights would require vast amounts of time and energy; and it is a fair assumption that intelligent aliens would explore the Galaxy using the cheapest and most efficient methods of exploration (at least, until they had reached that hypothetical level of engineering expertise at which faster-than-light travel became easy and cheap).

Bracewell suggested that a messenger probe could be here in our Solar System right now, trying to initiate contact with us. We may speculate endlessly about what such a messenger might look like; but it seems reasonable to suppose that, since it would be required to wait for millions of years before beginning its task, it would be heavily armoured against the effects of radiation and asteroid impacts (one particular danger would come not from gigantic floating mountains, but rather from micro-asteroids no bigger than a grain of sand, and yet capable of causing enormous damage over long periods of time). The probe would detect narrow-band signals (which do not occur in nature, and thus imply the presence of a manufactured transmitter), and then simply re-transmit them back to the planet. If the senders then repeated the transmission, the probe would know it had been recognised, and would then begin communication. This sounds straightforward enough; but the likelihood of such communication ever occurring depends upon the answers to two crucial questions.

Are there any other civilisations in our Galaxy?

And if so, how many?

'A LAND FAR AWAY . . .'

Born in Chicago in 1930, Frank Drake was the first person in history to use a telescope to listen for radio transmissions from other civilisations in space. His fascination with science began at an early age, and he and his childhood friends would spend long hours tinkering and experimenting with motors, radios and chemistry sets. Drake also developed an intense interest in astronomy, marvelling at the scale and grandeur of the Universe, and wondering whether it was possible that life existed on other worlds out there in that limitless celestial ocean.

After high school, he took the electronics course at Cornell University, where he also continued with his astronomical studies. He had felt uncomfortable bringing up the subject of extraterrestrial life with his parents, whose religious convictions ran counter to the idea. However, the atmosphere of intellectual freedom at Cornell dispelled all such qualms.

In 1951, during his junior year, Drake attended a lecture by Otto Struve, one of the world's leading astrophysicists. Towards the end of the lecture, Struve showed that there was mounting evidence that approximately half of all the stars in our Galaxy possess planetary systems, and went on to state that life could exist on some of those planets. At last, Drake had found someone who shared his belief that life could exist elsewhere in the Universe.

Following a stint in the US Navy, during which he served as electronics officer on the USS *Albany*, Drake attended Harvard graduate school with the intention of studying optical astronomy. However, the only summer position available was in radio astronomy. Because of his electronics experience in the Navy, Drake found himself well suited to the position: the radio astronomy equipment was in constant need of fine-tuning and repair. It was in this field that Drake subsequently decided to make his career.

On finishing graduate school in 1958, he took a position at the newly founded National Radio Astronomy Observatory (NRAO) in Green Bank, West Virginia. He began his search for extraterrestrial signals in 1960, using the observatory's 26-metre dish. Drake named the search 'Project Ozma', after the princess in L. Frank Baum's children's story *Ozma of Oz*. Drake later wrote: 'That book was part of the series that started with *The Wonderful Wizard of Oz*, which I had adored as a child. Like Baum, I, too, was dreaming of a land far away, peopled by strange and exotic beings.'

The initial search was a two-week observation of the stars Tau Ceti and Epsilon Eridani. After an uneventful observation of Tau Ceti, Drake turned his attention to Epsilon Eridani, and experienced the first of SETI's false alarms. 'A few minutes went by. And then it happened. WHAM! We heard bursts of noise coming out of the loudspeaker eight times a second.' Drake and his colleagues looked at each other in astonishment: had they really detected an alien signal so quickly, after observing just two stars out of the hundreds of millions in the Galaxy? While they had absolute confidence that they would one day succeed, they had expected it to take a lot longer than this! In fact, at this stage, they had not even planned a course of action in case a clear signal was received. The question that went around the room was: what do we do now?

The most obvious course of action was to check that the signal was not coming from Earth, and so they moved the telescope away from the star to see if the signal went away. They tried it and, sure enough, the signal disappeared. But when they pointed the telescope back at Epsilon Eridani, the signal did not come back. For the next week they trained the telescope on the star every day, in the hope that they might detect the signal again.

In the meantime, one of the telescope operators telephoned a friend in Ohio to share the news. That person in turn informed his local newspaper, and the press quickly picked up the story. Drake and his team were then deluged with enquiries about the mysterious signal. Although they tried to answer them as best they could, they knew so little about it themselves that their answers sounded like evasions:

'Have you really detected an alien civilisation?'
'We're not sure. There's no way to know.'
'But you did receive a message?'
'We heard a signal. We don't know what it was.'
'When will you know?'
'We can't say. It's hard to tell.'

From then on, the conspiratorially-minded became convinced that Project Ozma had received signals from another world, and that some sinister government agency had immediately sworn everyone involved to absolute secrecy. The reality, however, was far more mundane: ten days after the initial detection, the signal returned, and Drake later found it to be emitted by a high-flying aircraft, possibly a U-2 spyplane.

In preparation for the first SETI conference, a three-day meeting held at the NRAO in 1961, Drake developed an equation with the potential to determine the number of advanced technological civilisations existing in our Galaxy. Now known as the 'Drake Equation', it is written thus:

$$N = R f_p n_e f_l f_i f_c L$$

Where the number (N) of detectable civilisations in space equals the rate (R) of star formation, times the fraction (f_p) of stars that form planets, times the number (n_e) of planets hospitable to life, times the fraction (f_l) of those planets where life actually emerges, times the fraction (f_i) of planets where life develops intelligence, times the fraction (f_c) of planets with intelligent creatures capable of interstellar communication, times the length of time (L) that such civilisations remain detectable.

Although Drake admits that he has no firm values for most of the factors in his equation, he points out that it is not as speculative as some critics have suggested, since 'each phenomenon it assumes to take place in the Universe is an event that has already taken place at least once'.

The science writer John Gribbin offers one interpretation of the Drake Equation which is astonishing in its implications for the possibility of communication with the inhabitants of other star systems. Starting from the assumption that there are about 100 billion stars in our Galaxy, Gribbin speculates that:

> If a third of all stars have planets, two planets in each planetary system are habitable, a third of these are actually occupied by life, and technological civilisation has arisen on just 1 per cent of those planets . . . then there are a billion planets in our Galaxy alone on which technological civilisation has appeared. How many are around today depends on how long they survive, but it still seems possible that there could be millions of technological civilisations around in the Galaxy today.

15

PROTECTORS FROM BEYOND

THE GHOSTS OF FLIGHT 401

At 9.20 p.m. Eastern Standard Time on the evening of 29 December 1972, Eastern Airlines Flight 401 took off from John F. Kennedy International Airport, New York, bound for Miami International Airport, Florida. The plane, a Lockheed L-1011 Tri-Star, should have landed at Miami at 11.45 p.m.; instead, it crashed in the Florida Everglades just under 19 miles from its destination, setting the tragic scene for one of the most astonishing and meticulously investigated cases in the history of the paranormal.

The Tri-Star's flight crew consisted of Captain Robert A. Loft, First Officer Albert J. Stockstill, and Second Officer and Flight Engineer Donald A. Repo. At 55, Loft was a seasoned captain who had joined Eastern Airlines in 1940, and had been a captain for 22 years. Stockstill, at 39, was the youngest member of the flight crew; following a career in the US Air Force, he had joined the airline a year earlier, in 1971. The 51-year-old Repo had joined Eastern as an aircraft mechanic; and, following his qualification as an engineer in 1955, he gained his Commercial Pilot Certificate in 1967.

The flight from JFK proceeded without a glitch and, as the Tri-Star approached Miami, Loft ordered the landing gear lowered. Stockstill complied, and watched the three green lights on the instrument panel. The first two lit up, confirming that the port and starboard landing gear had deployed; the third representing the nose gear, however, remained unlit.

The flight crew's words were captured by the flight deck voice recorder, which was later retrieved undamaged from the wreckage. Captain Loft radioed Miami Air Traffic Control. 'Ah, tower – this is Eastern, ah, four zero one. It looks like we're gonna have to circle. We don't have a light on our nose gear yet.'

The tower responded: 'Eastern four oh one heavy: roger. Pull up, climb straight ahead to two thousand, go back to approach control.'

Loft did as requested, reporting back to the tower when he had reached the specified altitude. The nose gear light remained unlit, and so the Captain instructed Stockstill, who was flying the plane, to engage the automatic pilot.

Loft said he was sure the nose gear was down. 'There's no way it couldn't be.'

Stockstill replied: 'The tests didn't show that the lights worked anyway. It's a faulty light.'

In that sentence lies the essence of the Flight 401 tragedy, for although the flight crew didn't know it for certain, Stockstill was absolutely right: the nose gear light *was* faulty; the nose gear *had* deployed properly. They could have landed without further incident – but they didn't.

They had to be absolutely sure that the gear was down. Captain Loft had two options: first, he could replace the nose gear lamp; second, he could send Don Repo down into the forward avionics bay, which was situated directly beneath the flight deck. There, the Second Officer/Flight Engineer could use a telescope-like device known as an optical sight, and would be able to see directly whether the nose gear was up or down.

Captain Loft decided to follow both options. While Repo prepared to descend into the avionics bay, Loft and Stockstill struggled to replace the nose gear light. They had difficulty getting the fiddly thing out of the instrument panel, and more difficulty inserting its replacement.

Loft to Stockstill: 'You got it sideways then . . . Naw, I don't think it'll fit. You gotta turn it one quarter to the left.'

Stockstill to Loft: 'Bob, this **** [word censored on tape transcript] just won't come out.'

At some point, someone accidentally (and unknowingly) nudged the aircraft's control column, which disengaged the autopilot. No one was now flying the plane, which immediately began steadily to lose altitude.

Meanwhile, Repo had gone down into the avionics bay and was looking for the optical sight, which he had trouble finding. Up on the flight deck, a chime sounded quietly, indicating the plane's loss of altitude, but in the confusion of their attempts to replace the nose gear light, neither Loft nor Stockstill heard it.

Just a couple of minutes later, Stockstill said: 'We did something to the altitude.'

'What?' said Loft.

'We're still at two thousand, right?' said Stockstill.

'Hey, what's happening here?' demanded Loft. 'What's –'

His voice was cut off as the Tri-Star hit the swampy ground of the Everglades at nearly 200 mph.

Thanks to the swift action of the United States Coast Guard, the 75 survivors out of 176 passengers and crew were rescued quickly from the swamp's clammy embrace. Captain Loft and First Officer Stockstill died in the crash; Second Officer Repo was pulled out of the wreckage alive, but he died later in hospital.

The tragedy was meticulously investigated, recommendations were made (one of which was to add a flashing light to the low-altitude warning chime on the flight decks of Tri-Stars), and the dreadful fate of Flight 401 was consigned to the annals of aviation history.

Or rather, it would have been, had not flight crews and passengers on other Eastern Airlines flights begun to report strange encounters with the apparent ghosts of Captain Robert Loft and Second Officer Don Repo.

The matter came to the attention of investigative journalist John G. Fuller (who had previously published *The Interrupted Journey*, the definitive account of the famous Barney and Betty Hill UFO abduction case). Fuller spent some considerable time interviewing witnesses who claimed to have encountered the deceased flight officers, and published his findings in a bestselling book entitled *The Ghost of Flight 401*.

In one case, a Tri-Star was en route from New York to Mexico City, when one of the female flight attendants, who was in the lower galley, glanced at a microwave oven, and saw in the glass of the oven's door the face of Don Repo.

Understandably startled, and unsure whether she could believe her eyes, she called a colleague to come and take a look. Her colleague confirmed that it was indeed Repo's face in the glass of the oven door. The attendants went up to the flight deck and asked the Flight Engineer to go with them to the lower galley. He did so, and he too saw the face of the dead Second Officer from Flight 401. As if this were not astounding enough, Repo actually spoke to them, saying: 'Watch out for fire on this airplane.'

The aircraft reached Mexico City without incident; however, shortly after take-off on the return journey, one of the engines developed a fault and a fire broke out. The pilot shut the engine down and managed to return to Mexico City and land safely.

This was just one of many strange incidents which were reported to Fuller, usually by members of flight crews who wanted to remain anonymous for fear of ridicule, or worse. Eastern Airlines was not anxious to be associated with ghostly events and, like aircrews who witness UFOs, those who encountered Repo and Loft knew better than to broadcast the fact.

Captain Loft was seen on several occasions, but he spoke only once – and on this occasion he did not appear. A Tri-Star was on its way from New York to Miami. As it flew over the Everglades, the public address system switched on and a voice announced that they would be landing shortly and asked passengers to fasten their seatbelts and extinguish their cigarettes (this was in the days when people were allowed to smoke on aeroplanes). A normal-enough procedure, until the flight crew realised that none of them had made the announcement and the PA system had been switched off. The voice had been that of Captain Loft.

On another occasion, a flight attendant was counting passengers on a flight prior to take-off, and noticed that the plane had an extra passenger: a man dressed in an Eastern Airlines captain's uniform, who was sitting in one of the first-class seats. Assuming that he was 'dead-heading' (hitching a ride on an aeroplane, a quite normal occurrence), she approached the man and asked if he would rather make the journey in the jump seat on the flight deck. He did not reply, so she asked him again (she later reported that he appeared to be confused, in a kind of daze). When he still said nothing, the attendant grew concerned and asked the aeroplane's captain to come aft and speak to him. When the captain saw the silent flight officer, he halted, stunned, and said: 'My God, it's Bob Loft!' At that moment, the flight officer vanished before their eyes.

On many occasions, an unexplained drop in temperature accompanied the apparitions, which seemed to be centred on one particular aircraft, a Tri-Star designated number 318. In one of the strangest incidents, a passenger noticed a hazy, luminous mass hovering over the starboard wingtip. When the mass touched the wingtip, the plane rolled a little to the right, and then recovered. The passenger called a flight attendant, who also witnessed the phenomenon; she called the Flight Engineer, who was at a loss to explain it and suggested that it was merely a cloud (even though it was behaving like no cloud any of them had ever seen). Several minutes later, the 'cloud' moved to the port wingtip and again began to touch it periodically; and once again, the plane rolled a little each time it did so.

It was later revealed that several components from the doomed Flight 401 were undamaged, and were recycled in other Tri-Stars. The rumour began to spread that the planes in which Loft and Repo appeared had been fitted with these components. In spite of their reluctance to give credence to the reports of ghostly activity on their aircraft, Eastern Airlines nevertheless decided to remove these components and replace them with new ones.

During the course of his investigation, John G. Fuller decided to attempt to make contact with Loft and Repo by means of a séance. He apparently succeeded in contacting Repo via a ouija-board. As is often the case with this method, the messages received were fragmentary. When asked how the crash of Flight 401 had occurred, the response was: RELEASED CONTROLS ACCIDENTALLY. The ouija-board then spelled out several messages of love for Repo's wife and daughter.

When Fuller asked Repo where he was, the cryptic response was: GPNE TO TO UN POSNTN. Confused, Fuller asked if Repo was on Earth, to which the response was: NO. Again, Fuller asked where the Second Officer was. This time the response was: INFUTE INFINITE.

The results of Fuller's investigation are at once intriguing and unsettling, and yet they constitute one of the best-documented and most widely experienced hauntings in the history of the paranormal, and as such offer encouraging evidence that some part of the human personality survives physical death. Was it true, as Fuller suggests, that Captain Loft and Second Officer Repo held themselves responsible for the crash of Flight 401, and that they had returned to the material world in an effort to make amends by protecting other aircraft?

Perhaps the answer lies in two other encounters between flight crews and Don Repo. On one occasion, a Flight Engineer was making pre-flight checks, when he turned and saw Repo sitting at the instrument panel. Repo said to him: 'You don't need to worry about the pre-flight, I've already done it.' On another occasion, a Tri-Star captain encountered Repo on the flight deck. Repo said to him: 'There will never be another crash of a Tri-Star . . . we will not let it happen.'

16

THE WIZARD OF PASADENA

THE STRANGE DOUBLE LIFE
OF JACK PARSONS

Jack Parsons was one of the founding fathers of American rocket science, a self-educated genius whose development of early rocket propulsion systems helped make NASA's Apollo moon flights possible, and who founded the company that now makes the solid-fuel boosters used on the Space Shuttle. So fundamental was his contribution to space exploration that he has a crater on the Moon named after him. He also helped to create the Jet Propulsion Laboratory (JPL) in Pasadena, California, which is at the very forefront of manned and unmanned interplanetary flight. In fact, there is even a joke in the aerospace community that 'JPL' actually stands for 'Jack Parsons Laboratory' or 'Jack Parsons Lives'.

Parsons was a scientific visionary, one of the handful of human beings who believed in humanity's potential as a spacefaring species, and who had the courage and the intelligence not only to see what the future might bring, but to take the necessarily radical and often dangerous steps to realise that future. And the danger was real: in 1952, at the age of 38, Parsons was killed in a mysterious explosion, the cause of which has never been satisfactorily explained.

Parsons was the epitome of the American pioneer, almost a cliché, in fact. Tall, well built, ruggedly handsome, irresistible to women, a man who made things happen, he was a living embodiment of the larger-than-life heroes of the pulp science fiction stories he loved so much. Looking at photographs of him, one can easily imagine him at the controls of some outlandish rocketship, with a nubile young assistant at his side, heading for Mars or Venus, or out into the great galactic wilderness.

For Jack Parsons, however, rocket science was far from the whole story: there was another side to his character, another range of interests that consumed him just as fully as chemistry and engineering. In addition to being the all-American pioneer in a brand new science, he was also the practitioner of an ancient one, that of ceremonial magic. In the unlikely setting of 1940s California, Parsons would attempt an occult rite known as the Babalon Working, whereby he would try to summon a powerful and dangerous entity and incarnate it into human form through the impregnation of a willing female partner.

The apparent contradiction of a modern scientist believing so fervently in the power of magic is deliciously ironic and intriguing, and Parsons' early life and career give some clues as to the course into the realms of the occult and supernatural his life would take.

EARLY LIFE

On 2 October 1914, Marvel H. Parsons and his wife, Ruth, had a baby boy whom they named Marvel Whiteside Parsons. A year later, Ruth divorced her husband on grounds of adultery, and began calling the boy 'John', although she never legally changed his name. Family and close friends would later come to call him 'Jack', while to the scientific community he would always be known as 'John'.

Parsons would later write that at this time his mother began to cultivate in him an intense hatred for the father he never knew. In his essay 'Analysis By A Master of the Temple', written when he was 34, Parsons wrote (bizarrely referring to himself in the second person):

> Your father separated from your mother in order that you might grow up with a hatred of authority and a spirit of revolution necessary to my work. The Oedipus complex was needed to formulate the love of witchcraft, which would lead you into magick, with the influence of your grandfather active to prevent too complete an identification with your mother.

In the absence of his father, the principal male influence in Parsons' early life was his grandfather, Walter. In his fascinating biography of Parsons, *Sex and Rockets*, John Carter speculates that it may have been Walter who encouraged the boy's early interest in rockets and fireworks.

Parsons was a lonely child with few friends; but, as he later wrote, this isolation helped to develop a love of literature, not to mention contempt for the human herd, which he was to find essential in his adult pursuits. In addition, he developed a powerful hatred of Christianity and what he called its 'guilt sense'.

When he was in the eighth grade at school, Parsons was saved from a bully's attentions by a boy named Edward Forman. The two became firm friends, and discovered that they had similar interests, such as the novels of Jules Verne, and Hugo Gernsback's newly founded science fiction magazine *Amazing Stories*. They also both loved rockets, and they quickly graduated from setting off fireworks in Parsons' back yard to experimenting with small, solid-fuel projectiles.

In 'Analysis By A Master of the Temple', Parsons wrote:

> Early adolescence continued the development of the necessary combinations. The awakening interest in chemistry and science prepared the counterbalance for the coming magical awakening, the means of obtaining prestige and livelihood in the formative period, and the scientific method necessary for my manifestation. The magical fiasco at the age of sixteen was needful to keep you away from magick until you were sufficiently matured.

The exact nature of this 'magical fiasco' is unknown. Around this time, his grandfather passed away, leaving the 17-year-old Parsons without a significant father figure and role model in his life. As John Carter notes: 'Most of his adult life [Parsons] sought out others to fulfil this role.'

Parsons and Forman continued to experiment with rockets, often at some risk to life and limb, and in 1932 Parsons went to work for the Hercules Powder Company of Pasadena. The following year he graduated from the private University School and, together with Forman, he attended the University of

Southern California. Neither of them graduated.

During their experiments, Parsons and Forman corresponded frequently with other scientists in the field, including Robert Goddard in Roswell, New Mexico, Willy Ley in Germany (Ley would later flee the Nazis and move to America, bringing claims that the Nazis were obsessed with the occult) and Hermann Oberth, the German rocket pioneer. Unfortunately for them, Parsons and Forman eventually realised that their correspondents seemed to be more interested in obtaining information from them than volunteering their own.

In the spring of 1935, Parsons married Helen Northrup, whom he had met at a church dance, and later he wrote in 'Analysis By A Master of the Temple': 'The early marriage to Helen served to break your family ties and effect a transference to her, away from a dangerous attachment to your mother.' Around this time, there seems to have been a downturn in the Parsons family's financial situation (until then, the family had been quite wealthy). That autumn, Parsons read an article in the *Pasadena Evening Post* about a lecture given at the California Institute of Technology (Caltech) concerning the rocket experiments of the Austrian Eugen Sänger. The lecture's concluding speculations on the possibility of 'stratospheric passenger carriers' or manned rocket ships reinforced Parsons' belief in the boundless potential of the radical propulsion systems with which he and Forman were experimenting.

GALCIT

The two men were keen to begin working with liquid-fuelled rockets, but lacked the funds to get the project going. They approached William Bollay, the graduate student who had delivered the lecture on Sänger's work, enquiring about the possibility of working at Caltech (more specifically, at GALCIT, the Guggenheim Aeronautical Laboratory, California Institute of Technology), which they hoped would finance their experiments. Bollay was not in a position to offer direct help, but he put them in touch with Frank Malina, who was working on his PhD at Caltech, and who was close to GALCIT's director, the Hungarian Professor Theodore von Kármán. Malina immediately recognised their talents; although without formal qualifications, they made an excellent team, Parsons the talented chemist, and Forman the expert engineer. Von Kármán agreed to allow the two mavericks to use the GALCIT facilities.

Funds were not quite as plentiful as they had hoped. As Carter notes, at that time 'rockets were still viewed as science fiction by the public. The mere mention of rockets had people thinking of such impossible things as trips to the moon and rarely anything else.' Nevertheless, von Kármán persuaded GALCIT to let Parsons, Forman and Malina use three acres of land, which Caltech leased from the city of Pasadena. The area was called the Arroyo Secco, and lay at the foot of the San Gabriel Mountains behind the Devil's Gate Dam. It is now the site of NASA's Jet Propulsion Laboratory.

The experiments began slowly, mainly due to the shortage of funds from GALCIT; Parsons and Forman had to take jobs to finance the tests they had planned, and to meet their basic living expenses. On one occasion, Malina wrote:

> Parsons and I drove all over Los Angeles – looking for high pressure tanks and meters. Didn't have any luck. Two instruments we need costs [sic] $60 a piece and we are trying to find them second hand. I am convinced it is a hopeless task. Will have to approach Kármán.

A few weeks later, the situation hadn't changed.

> The early part of this week, Jack Parsons and I covered much of Los Angeles looking for equipment. Our next lead points to Long Beach. Parsons is planning to start manufacturing explosives with another fellow [Forman]. Up to the present time he has been working for an explosives concern. Hope they make a go of it. I have found in Parsons and his wife a pair of good intelligent friends.

In April 1937, Parsons, Malina and Forman left the Arroyo Secco and moved to one of the GALCIT laboratory buildings on the Caltech campus in Pasadena. They had the ever-sympathetic von Kármán to thank. The following month, the group was joined by Hsue-shen Tsien from China. Shortly afterwards, a meteorology student named Weld Arnold presented the group with a small package wrapped in newspaper. Inside was $1,000 in small bills. No one thought it prudent to ask him how he had come by this impressive amount of cash. In recognition of his contribution, Arnold was appointed unofficial photographer.

The group's time on the Caltech campus did not begin well, as Carter writes in his biography of Parsons:

> Campus residents soon resented the group's presence. Testing was loud and violent. Immediately after their arrival on campus, the group met with their first disaster. In the lab they had mounted a rocket motor to a 50-foot pendulum. They measured the swing of the pendulum to calculate the thrust produced. The first motor tested exploded, filling the building with a cloud of methyl alcohol and nitrogen dioxide. A thin layer of rust formed on many pieces of valuable equipment throughout the lab. The other campus residents started referring to the group as 'the Suicide Squad'. Parsons would remember this event four years later, when he finally figured out a way to put the volatility of red fuming nitric acid (RFNA) to good use.

Von Kármán decided that it would be better for the group to move their experiments outside the laboratory. They rebuilt the pendulum, making it five times stronger than was considered necessary, on von Kármán's orders. Two years later, another explosion destroyed the strengthened pendulum, and sent a lump of metal hurtling into a wall where Malina had been standing moments earlier. In spite of his sympathetic attitude towards their work, von Kármán decided there was nothing for it but to send them back to the Arroyo Secco.

These setbacks notwithstanding, the group continued their research undaunted, and they were rewarded in the autumn of 1938 when the National Academy of Science (NAS) Committee on Army Air Corps Research expressed an interest in using rockets to assist the take-off of heavily laden aircraft. The NAS awarded Caltech a $10,000 contract to develop rocket propulsion with the aim of providing 'super-performance' for propeller-driven aircraft.

The tests that followed resulted in the development in August

1941 of the JATO unit (JATO stands for Jet Assisted Take Off, the word 'jet' replacing 'rocket' because of the science fiction connotations of that word). The JATO units had to be used almost immediately; they could not be stored for long periods of time or in extremes of temperature, otherwise they would become unstable and explode. According to Carter: '[Parsons] and Forman would get up very early and prepare the JATOs, nap a little, then meet the others at March Field [the testing site]', and von Kármán wrote: '[Parsons] used a paper-lined cylinder into which he pressed a black-powder propellant of his own composition in one-inch layers'. In August, the group was ready to test Parsons' JATOs on actual aircraft, the first time this type of rocket had ever been used.

The aircraft they chose for the first test was the Ercoupe, a mail-order hobby aircraft, which was small, light and difficult to stall. The group fitted 12 JATOs under the Ercoupe's wings; each motor delivered 28 pounds of thrust during its 12-second burn. The plane's propeller was removed for the test flight. The pilot, Captain Homer A. Boushey Jr of the Army Air Corps, took the controls and soared into the air, demonstrating the viability of a propulsion technology that would ultimately take human beings to the Moon.

A NEW DIRECTION

As the 1930s drew to a close, Parsons made a discovery that would, as Carter writes, 'change the direction of [his] personal life as much as Bollay's lecture changed the direction of his professional life'. He was looking through the library of an old friend, Robert Rypinski. The two had met years before, when Rypinski had sold Parsons a used car. It was among his friend's collection of books that Parsons came across a copy of Aleister Crowley's *Konx Om Pax*, originally published in 1907. The book had proved too complex and abstruse for Rypinski, so he gave it to Parsons, who almost immediately began to correspond with Crowley. Rypinski later said that to Parsons the book was 'like real water to a thirsty man'.

Aleister Crowley had visited Los Angeles briefly in 1915, and seems to have been singularly unimpressed with the place. It was not until 1935 that Crowley's magical organisation, the Ordo Templi Orientis (OTO), had an official representative in the city, in the person of an expatriate Englishman named Wilfred Talbot Smith. Hailing from Tonbridge, Kent, Smith had been an associate of Charles Stansfeld Jones (Frater Achad, or Brother Unity), founder of the Agape Lodge in Vancouver, the first OTO lodge on the North American continent. Although Smith had risen to a position of authority within the Agape Lodge, and had met Crowley there in 1915, he fell out of favour and was later expelled. He moved to Los Angeles in 1930, and opened a new Agape Lodge there, which gave weekly performances of Crowley's Gnostic Mass. As Carter explains:

> The Gnostic Mass was Crowley's replacement for the 'corrupted' mass celebrated by the Roman Catholic and Eastern Orthodox churches. The text of the Mass, referred to as Liber XV, was written in Moscow in 1913. Partly inspired by the Russian Orthodox Mass, it is surprisingly Christian in its symbology, and there is nothing obscene about it despite what certain detractors may say. In fact, it is reminiscent of Wagner's *Parsifal* with its repeated references to 'lance' and 'Grail', which are its most suggestive elements. As in the Christian Mass, the Holy Spirit is invoked often. The most openly erotic element in the Mass occurs when the priestess disrobes for part of the ceremony.

In 1939 a scientist colleague of Parsons (whose identity is unknown) took him to Smith's house in Hollywood, where they participated in the Gnostic Mass. Thereafter, Parsons attended regularly with Helen. His meeting with Smith was immensely important to Parsons: not only had he discovered an outlet for his powerful sense of mystery and romance, he had also found the father figure he had lost when his grandfather died.

At this time, the Agape Lodge was run by Smith and his mistress, Regina Kahl, both of whom were extremely authoritarian. In the celebrations of the Gnostic Mass, Smith took the role of Priest and Kahl that of Priestess. Parsons and Helen were initiated into the Lodge on 15 February 1941 and, like many other recruits, they simultaneously became members of Crowley's magical order, the Argenteum Astrum. From the beginning, the handsome and dynamic Parsons made an impression on his fellow members, including the actress Jane Wolfe, who had appeared in several silent films, and who had spent time with Crowley at his Sacred Abbey of Thelema in Sicily. The following is from her *Magical Record*, and was written in December 1940:

> Unknown to me, John Whiteside Parsons, a newcomer, began astral travels. This knowledge decided Regina to undertake similar work. All of which I learned after making my own decision. So the time must be propitious . . . 26 years of age, 6' 2", vital, potentially bisexual at the very least, University of the State of California and Cal Tech, now engaged in Cal Tech chemical laboratories developing 'bigger and better' explosives for Uncle Sam. Travels under sealed orders from the government. Writes poetry – 'sensuous only,' he says. Lover of music, which he seems to know thoroughly. I see him as the real successor of Therion [Crowley]. Passionate; and has made the vilest analyses result in a species of exultation after the event. Has had mystical experiences, which gave him a sense of equality all round, although he is hierarchical in feeling and in the established order.

Parsons' magical name was Frater TOPAN (the initials stood for *Thelemum Obtentum Procedero Amoris Nuptiae* ('the obtainment of *thelema* |will] through the nuptials of love'). Helen became Soror Grimaud.

Jane Wolfe's comment regarding Parsons' potential bisexuality is probably inaccurate. Carter notes that this may have occurred to her as a result of Parsons' closeness to Smith, coupled with the fact that Parsons sweated profusely and attempted to mask his strong body odour with heavy cologne. 'A man who wore a lot of cologne at the same time that he displayed an above-average interest in another man may have inadvertently given the impression of bisexuality.'

Whatever Parsons' sexual proclivities may have been, his attitude to sex itself was clear enough. His enormous house at 1003 South Orange Grove Avenue in Pasadena became home to an astonishing variety of free-living and free-loving types, including musicians, artists, writers and anarchists. It was not long before Parsons' strange, eclectic and ever-changing group of tenants became the subject of rumour and lurid speculation in the neighbourhood. There were numerous whisperings of black magic rituals and sexual orgies, and one evening in 1942 several people called the police, claiming that a pregnant woman was jumping naked through a fire in Parsons' back yard. When several officers arrived to investigate, Parsons politely informed them that as a reputable scientist he had no interest in such bizarre silliness. The police were convinced, and left without taking the matter further.

In keeping with his scientific and occult interests, Parsons counted many science fiction writers among his friends and acquaintances. During the 1940s, science fiction was still in what is known as its 'golden age', and the West Coast boasted a number of key writers. Many of these gathered at Parsons' home, including such science fiction luminaries as Jack Williamson, A.E. van Vogt, Robert Heinlein, Ray Bradbury and science fiction historian Forrest J. Ackerman. In addition to being influenced by his reading of science fiction, Parsons seems also to have influenced at least one of these writers: Robert Heinlein used some of Parsons' ideas in what would become one of the most famous of all science fiction novels, *Stranger in a Strange Land*.

THE BABALON WORKING

In August 1945, L. Ron Hubbard (who would later find fame as the inventor of Scientology) came into Parsons' life. The two were introduced while Hubbard was on leave from the Navy, and they immediately found that they had much in common, most notably their interest in science fiction and the mysteries of the occult. Parsons was greatly impressed with Hubbard, and invited him to move into the house on South Orange Grove Avenue. However, it was not long before Hubbard began a passionate affair with Parsons' girlfriend, Betty Northrup, who was Helen's younger sister (Helen had already left Parsons for Wilfred Smith). In a letter to Aleister Crowley, Parsons noted that, although he was not a formally trained magician, Hubbard nevertheless possessed an impressive understanding of the subject. Parsons believed that Hubbard was in direct contact with some transhuman intelligence, perhaps his Holy Guardian Angel. He added that he needed a magical partner for the many experiments he was planning.

The most important of these experiments was known as the Babalon Working, by which Parsons hoped to create an elemental being, a supernatural intelligence, the creation of life signifying the magician's power over nature. This operation is far too long and complex to describe in its entirety, and so we must limit ourselves to a brief description. Together with Hubbard, Parsons began the Babalon Working in January 1946, having consecrated his magical equipment, including the Air Dagger, a magical weapon used to provide a focus for occult power. Choosing one of the squares from the Enochian Air Tablet (Enochian was the alleged angelic language discovered by the great Elizabethan mathematician and occultist John Dee), Parsons copied its magical symbols related to the element of Air onto virgin parchment. The symbols consisted of a planetary sign, a zodiacal sign, a certain permutation of the four signs of the elements, and an Enochian letter.

Many who knew Hubbard were of the opinion that he was something of an eccentric confidence trickster, and John Carter makes an interesting point that Hubbard's relationship with the good-hearted and diligent Parsons was akin to the dubious Edward Kelly's relationship with John Dee.

The first task of the Working was to trace in the air with the magical dagger the sign of the pentagram. Next, Parsons recited the Invocation of the Bornless One from the grimoire known as the *Lesser Key of Solomon*. After reciting the Third Call, which is designed to summon EXARP, the Angel of the Air Tablet, followed by an Invocation of God, Parsons then invoked the Six Seniors of the Air Tablet, whose names are HABIORO, AAOZXAIF, HTNMORDA, AHAOZAPI, AVTOTAR and HIPOTGA. Next, he conducted an Invocation of the Wand, in which he fertilised the parchment containing the symbols from the Air Tablet by masturbating over it. When the first part of the Working was over, Parsons performed the necessary banishing rituals, including the Licence to Depart from the *Lesser Key of Solomon*.

While performing this part of the Working, Parsons noted that a powerful windstorm had begun, in keeping with his conjuration of the Angel of the Air Tablet. The wind continued as he performed the invocation again, this time playing Prokofiev's Violin Concerto No. 2 as he did so.

On the night of 10 January, he was awakened at midnight by nine loud knocks. When he got up to investigate, he noticed a lamp lying smashed on the floor. This was not an intended result of the Working, and meant that magical energy had been somehow misdirected. Parsons was an experienced enough magician to realise that something was wrong, but he continued with the Babalon Working anyway. (Carter notes an interesting coincidence here: the Enochian word HUBARD means 'living lamps'.)

On 14 January, as Parsons began the Working for that day, there was an unexpected power cut. According to Parsons, 'another magician who had been staying in the house and studying with me was carrying a candle across the kitchen when he was struck strongly on the right shoulder, and the candle knocked out of his hand'.

The following day, Parsons noted that Hubbard had developed a form of astral vision, with which he had described an old enemy of Parsons'. Later, in his room, Parsons became aware of a strange, buzzing, metallic voice, which demanded to be allowed to go free. (The 'old enemy' could have been a corrupt police officer, Captain Earle E. Kynette. In May 1938, Parsons had been called as an expert witness in the trial of Kynette, who had been accused of planting a pipe bomb in the car of a vice investigator named Harry Raymond. Parsons' testimony had led to Kynette's conviction for murder, in spite of death threats made against him.)

Parsons performed the Licence to Depart; however, although the spirit returned to its rightful place, the strange feeling of oppressive tension continued for several days. The only apparent occult manifestation resulting from his efforts thus far had been the windstorm, and Parsons was deeply disappointed with his magical work. However, by 23 February 1946, things had changed very much for the better. He excitedly wrote to Crowley that he had finally found his elemental, in the form of a woman who arrived one evening following the conclusion of the operation. He described her as having red hair and green eyes, just as he had specified in the ritual.

Parsons' 'elemental' was, in fact, Marjorie Elizabeth Cameron, a strong-minded, self-reliant artist who had recently arrived at the Agape Lodge. Although she was unaware of the nature of the Babalon Working, she participated in its next phase, for which Parsons made her a protective talisman. Parsons was more than satisfied with the woman he saw as his elemental, writing of her (as usual, in the second person) that she 'demonstrated the nature of woman to you in such unequivocal terms that you should have no further room for illusion on the subject'.

Parsons and Cameron became inseparable; she became his magical partner, and he educated her in the mysteries of the occult. In late February, Cameron went back to New York briefly, in order to finish with her boyfriend. When she returned to Pasadena, she discovered that she was pregnant by Parsons, who in the meantime had gone into the Mojave Desert to perform a magical rite the nature of which is unclear, but which resulted in a revelation described in his book *Liber 49*.

In March 1946, the Babalon Working entered its second phase, which included material Parsons had received while in the Mojave Desert. Hubbard, who had been away for several days, returned, saying that he had had a strange vision of a 'savage and beautiful woman riding naked on a great, cat-like beast'. He was anxious to give Parsons a message, and so they made the necessary magical preparations, constructing an altar and dressing in robes, Hubbard in white and Parsons in black. Hubbard suggested that they play Rachmaninov's 'Isle of the Dead' as background music. At 8.00 p.m., Hubbard began to dictate, with Parsons transcribing as he spoke.

The material received represented a triumphant contact with the goddess Babalon, in which she gave them detailed instructions for the next phase of the Babalon Working, which Parsons conducted with Marjorie Cameron.

As the Working progressed through its many phases, Parsons wrote to Crowley to keep him informed of how it was going. Although Parsons gave the impression that he was well in control, Crowley became concerned, and on 15 March he wrote to Parsons warning him about his relationship with Cameron. Crowley reminded him of the advice given by the great magical theoretician Eliphas Lévi, that a Magus may love his elementals too much, and that this has the potential to destroy him.

Crowley had an accurate suspicion of Parsons' ultimate goal, for he wrote: 'Apparently he, or Hubbard, or somebody, is producing a moonchild. I get fairly frantic when I contemplate the idiocy of these goats.'

With the Babalon Working, Parsons was indeed trying to create a moonchild, an elemental being into whom Babalon would be able to incarnate. It is unlikely, however, that he was attempting to cause a moonchild to be born. As Carter notes, Cameron later claimed that she had become pregnant by Parsons, and had had an abortion with Parsons' agreement. If the couple had been trying to create a moonchild, Cameron would not have aborted her baby. Carter suggests that Parsons was hoping that an adult female would arrive in the same way as had Cameron. In the arrival of the moonchild, Parsons expected and hoped for the arrival of a female messiah, in the form of the incarnated goddess Babalon.

In February 1946, while undertaking the Babalon Working, Parsons formed a company with Hubbard and Betty, called Allied Enterprises. They intended to buy boats on the East Coast, transport them to California and resell them at a profit. Hubbard invested nearly $2,000, Parsons nearly $21,000.

After the Babalon Working had been completed, Hubbard left with Betty and $10,000 of the company's money. The plan had been for them to go to Miami, buy a boat and sail it to California for resale; however, Parsons discovered that they had stayed in Miami, and had no intention of selling the three boats they had bought. He went to Miami and Howard Bond's Yacht Harbour, where he was informed that Hubbard and Betty had taken one of the boats out to sea. Furious, Parsons returned to his hotel room, where he summoned Bartzabel, a warlike spirit associated with Mars, to go after them. When this operation had been completed, Hubbard's vessel was struck by a sudden squall at sea, and only just made it back to Miami in one piece. Parsons then filed a lawsuit against Hubbard, and got two of the boats back, along with a promissory note for nearly $3,000. Parsons returned to California, but this was the last he ever heard from Hubbard and Betty.

At this time, Parsons was also under investigation by the FBI for several reasons, including his membership in occult groups and his suspected communist sympathies. In addition, the FBI found that while working for Hughes Aircraft, Parsons had taken several research files. Parsons was highly sympathetic to the state of Israel, and it has been suggested that he may have been included in a plan to secure a nuclear weapon for the newly founded state. As a result, Parsons permanently lost his security clearance, and was reduced to working at a gas filling station and occasionally designing explosive special effects for the film industry.

It seems that on Hallowe'en 1948, Babalon returned to Parsons and instructed him to resume his magical activities. This he did by taking the Oath of the Abyss in order to unite himself with the Universal Consciousness through denial of the material world. During the ceremony, he took the magical name Belarion Armiluss Al Dajjal, AntiChrist.

On Tuesday 17 June 1952, Parsons was working on an unknown experiment in his garage. At 5.45 p.m., two explosions ripped through the building. Parsons was killed instantly. There is a great deal of mystery and controversy surrounding his violent end. Some believe that the explosions were caused by Parsons dropping a container of fulminate of mercury, although, considering his experience with explosives, this does not seem plausible. Others have suggested that he was murdered by crooked colleagues of Earle Kynette, the corrupt police officer against whom Parsons had testified in 1938. There is also the theory that Parsons was engaged in research regarding the creation of a homunculus, an artificial man said by alchemists to possess magical powers.

The magical records left behind by Parsons have enabled (as was his plan) other occultists to evaluate the work upon which he was engaged, the intention of which was to understand the nature of reality and humanity's place in the Universe.

One of the most bizarre yet thought-provoking (not to mention highly debatable) interpretations of Parsons' work was put forward by the British occultist Kenneth Grant, who suggests that the Babalon Working was phenomenally successful, although not in the way Parsons had expected. Noting that the Working was completed just months before 'the wave of unexplained aerial phenomena now recalled as the Great Flying Saucer Flap', Grant suggests that the effect of the Babalon Working was to open a gateway through which 'something' was able to gain entry into our world . . .

17

SLEEPING SATELLITE

THE MYSTERY OF THE MOON'S ORIGIN

In Chapter 10, we looked at the controversy surrounding anomalous structures photographed on the Moon, and the suggestion by some researchers that they may be artefacts left behind by an extraterrestrial civilisation in the distant past. Astonishing as this claim is, it is as nothing compared to another theory which has been circulating for many years, a theory so bizarre and outlandish that it strains credulity to the breaking point and beyond.

Simply put, theory states that what we know as the Moon, that familiar, beautiful orb that has brightened the nights of every man, woman and child on Earth since the dawn of humanity, is not what we think it is. It is not a natural celestial object. It is artificial.

It is a gigantic alien spacecraft.

How could anyone come to such a conclusion? It is one thing to suggest that curious, unidentified lights flitting through the night sky might be the products of an unknown, non-human technology. But to suggest that the Earth's natural satellite, a sphere 3,476 kilometres in diameter, with a surface area of nearly 40 million square kilometres, is actually the ultimate UFO . . . well, to put it mildly, surely they're wrong!

Before examining the 'Spaceship Moon' theory in greater detail, it is worth pausing to acquaint ourselves with some basic facts about our nearest planetary neighbour.

In spite of its brightness in the night sky, the Moon is actually not a particularly reflective body: in fact, it only reflects about 7 per cent of the sunlight falling upon its surface. It orbits the Earth at a mean distance of 384,402 kilometres – 100 times closer than any satellite of any other planet in the Solar System. The eccentricity of its orbit means that its distance varies from 356,410 to 406,740 kilometres, which causes its apparent size when seen from Earth to vary by as much as 10 per cent.

The Moon is tidally locked to Earth (this is also known as 'captured rotation'), which means that it is always presenting the same face to us, and that the far side is forever hidden from direct view (it is a mistake to speak of the 'dark side' of the Moon: each side receives two weeks of sunlight each month). Although it is relatively small as planetary bodies go in our Solar System, it is nevertheless quite huge in relation to the Earth: its mass is 1/81 that of our planet. The next largest mass ratio between a major planet and its satellite is that of Titan and Saturn, and Titan's mass is only 1/4175 that of its parent world. The Moon's size in relation to Earth has led many scientists (including the great science fiction writer and science populariser Isaac Asimov) to suggest that the Earth–Moon system should be considered a double planet.

The Moon possesses low mass and density, so that the gravity at its surface is only about one-sixth that at sea-level on Earth. This means that any atmosphere it might have had in the distant past has long since drifted into space. As a result of the near-vacuum on the lunar surface, the moonscape is not subject to the same processes of erosion as the Earth, such as through wind and chemical action. The principal method of erosion on the Moon's surface is freeze–thaw action caused by the massive fluctuations in temperature. The maximum temperature during daytime is 130°C, and at night the temperature falls to –160°C. Over long periods of time, the resulting thermal stresses fracture rocks, turning them to dust over many millions of years.

In addition, constant bombardment by meteorites, most of which are extremely small (since the Moon has no atmosphere to shield it as the Earth does – there are no 'shooting stars' on the Moon) results in further erosion of the surface features at all scales. This accounts for the smooth and gently undulating landscape, so different from the towering peaks and razor-sharp crags envisaged by early science fiction writers and artists (even the magnificent *2001: A Space Odyssey*, so convincing in every other respect, made this mistake). There is very little seismic activity on the Moon: there are only about 3,000 moonquakes per year, almost all of which are below magnitude 3 on the Richter Scale, compared with about 800,000 earthquakes of magnitude 4 and below here on Earth.

The lunar surface is divided (in basic terms) between highlands and the so-called *maria*, or 'seas', the flat areas that give the lunar disc the appearance of a face (the 'man in the moon'), and which are the result of ancient outpourings of magma onto the surface.

COMPETING THEORIES OF THE MOON'S ORIGIN

Now that we have familiarised ourselves a little with the facts of the Moon's basic nature, we can look at the four models of lunar formation (there are four because no one is sure exactly how the Moon was created). They are: fission, binary accretion, capture and giant Earth impact.

According to the fission theory, the Earth and Moon broke apart into two unequal masses in the very early stages of the Earth's formation about 4.6 billion years ago. However, radiometric data gathered by the Apollo missions imply that had the Earth and Moon originated from the same primordial body, they would have cooled at exactly the same rate following separation, which is physically impossible.

In the binary accretion model, the Moon formed from a ring of debris orbiting the early Earth, and which failed to be incorporated into the body of our planet. However, this model also has problems: for one thing, the oxygen isotope ratios for the Earth and Moon are too similar for the bodies to have a separate accretion history.

The capture theory suggests that the Moon formed completely independently from the Earth, probably elsewhere in the Solar System but possibly in interstellar space, and later encountered the Earth and was captured by the larger planet. Once again, however, this theory has problems, the most serious of which is the fact that, while it is easy to account for the *departure* of an orbiting body in dynamical terms, it is much more difficult for a body to be *captured* and then to settle into a near-circular orbit, such as that followed by the Moon.

Finally, the giant Earth impact theory is the one that most astronomers favour at the moment. According to this theory, during the process of formation, there was a gargantuan collision between the early Earth and another proto-planet the size of Mars. So catastrophic was this impact that the Earth was very nearly destroyed, and a massive amount of material was thrown off into space. The Moon was thus created from a cloud of material from the proto-Earth, combined with that of the impacting proto-planet.

MYSTERIES OF THE MOON

We tend to think that we know pretty much everything about the Moon; after all, how much *is* there to know about a big ball of rock and dust? The fact is, though, we have set foot on the Moon a grand total of six times, and while a great deal of good science was done there, and with the 842 pounds of rocks and soil samples that the Apollo astronauts brought back with them, the Moon is still a very mysterious object.

Dr Robert Jastrow, who was the first chairman of NASA's Lunar Exploration Committee, called the Moon 'the Rosetta Stone of the planets'. He added: 'The Moon is more complicated than anyone expected; it is not simply a kind of billiard ball frozen in space and time, as many scientists had believed. Few of the fundamental questions have been answered, but the Apollo rocks and recordings have spawned a score of mysteries, a few truly breath-stopping.'

One of the most intriguing things about the Moon is the so-called 'mascons' (short for 'mass concentrations'), which are large, circular masses lying 20 to 40 miles beneath the surfaces of the *maria*. The mascons were first detected because the gravitational effects of their large mass distorted the orbits of spacecraft flying over them. Some scientists have suggested that these objects are heavy nickel-iron meteorites which ploughed into the Moon in its early history, while others (not all of whom are professional scientists, it must be said) claim that the mascons' circular shape and relatively shallow depth beneath the lunar surface hint at another origin altogether. According to Don Wilson, who has been an advocate of the 'Spaceship

Moon' theory for many years, says: 'It now appears that the mascons are broad, disk-shaped objects that could be possibly some kind of artificial construction. For huge circular disks are not likely to be beneath [the] *maria*, centered like bull's-eyes in the middle of each, by coincidence or accident.'

It has been known for many years that the Moon is less dense, on average, than the Earth. In the February 1962 issue of the journal *Astronautics*, the NASA scientist Dr Gordon MacDonald wrote: 'If the astronomical data are reduced, it is found that the data require that the interior of the Moon be less dense than the outer parts. Indeed, it would seem that the Moon is more like a hollow than a homogeneous sphere.' However, being a properly cautious scientist, he added: 'This suggests that there are inconsistencies either in the reduction of the observations of the Moon's motion or in the numerical development of the lunar theory.'

Since natural satellites cannot be hollow, this is an astonishing suggestion (and Dr MacDonald was quite right to qualify it by adding that something appeared to be wrong, either with the data or the observations).

Further evidence that the Moon's interior is quite different from what one would expect from currently understood models of planetary formation came to light on 20 November 1969. Having completed their lunar excursion, the astronauts of *Apollo 12* returned to the command-service module in orbit, and then jettisoned the lunar module ascent stage and sent it crashing into the lunar surface, about 40 miles from the landing site. During their excursion, the astronauts had set up highly sensitive seismic detectors, which would yield valuable information about the Moon's internal structure following the impact of the lunar module.

The spacecraft duly smashed into the Moon, but to say that the results of the experiment were unexpected would be an understatement. Following the impact, the Moon reverberated like a bell for half an hour. Scientists at NASA were perplexed to say the least. Dr Frank Press of MIT was quoted as saying that the reverberations were 'quite beyond the range of our experience'.

The experiment was repeated during the ill-fated Apollo 13 mission, when the launch vehicle's third stage was sent crashing into the Moon. A Saturn V third stage is a lot bigger than a lunar module, and this time the man-made object impacted on the lunar surface with the equivalent of 11 tons of TNT. The result was even more extraordinary: the reverberations lasted for more than three hours.

VASIN AND SHCHERBAKOV

The most comprehensive expression of the 'Spaceship Moon' theory was published by two members of the Soviet Academy of Sciences, Mikhail Vasin and Alexander Shcherbakov, in the journal *Sputnik*. Whether or not one agrees with its premise or conclusions, the article, entitled 'Is the Moon the Creation of Intelligence?', remains a tour de force of imaginative speculation.

In their introduction to their article, Vasin and Shcherbakov write that no one seems to have looked at the Moon in quite the same way as Mars, in an attempt to detect signs of large-scale engineering projects. They imply that the lunar surface has features which are just as puzzling and intriguing as the 'canals' of Mars, which had caused such controversy at the turn of the twentieth century when Percival Lowell made his mistaken claim that Mars was home to a dying race (see Chapter 10). Vasin and Shcherbakov add, almost apologetically, it seems, that their intention is to abandon 'the traditional paths of "common sense"', and enter the realm of 'what may at first sight seem to be unbridled and irresponsible fantasy.'

They suggest that the Moon is, in fact, a hollow sphere: a gigantic alien spacecraft, the interior of which contains everything required for the functioning of such an artefact, including fuel, engines, living quarters, navigational equipment, and so on. They describe the Moon as a 'caravel of the Universe', a 'Noah's Ark of intelligence', capable of supporting an entire civilisation on a voyage through space that might last thousands or even millions of years.

Under the subtitle 'A Noah's Ark?' the Soviet scientists write:

> If you are going to launch an artificial sputnik, then it is advisable to make it hollow. At the same time it

157

would be naïve to imagine that anyone capable of such a tremendous space project would be satisfied simply with some kind of giant empty trunk hurled into a near-Earth trajectory.

It is more likely that what we have here is a very ancient spaceship, the interior of which was filled with fuel for the engines, materials and appliances for repair work, navigation instruments, observation equipment and all manner of machinery . . . in other words, everything necessary to enable this 'caravel of the Universe' to serve as a kind of Noah's Ark of intelligence, perhaps even as the home of a whole civilisation envisaging a prolonged (thousands or millions of years) existence and long wanderings through space . . .

Naturally, the hull of such a spaceship must be super-tough in order to stand up to the blows of meteorites and sharp fluctuations between extreme heat and extreme cold. Probably the shell is a double-layered affair – the basis a dense armouring of about 20 miles in thickness, and outside it some kind of more loosely packed covering (a thinner layer – averaging about three miles).

Since the Moon's diameter is 2,162 miles, then looked at from our point of view it is a thin-walled sphere. And, understandably, not an empty one. There could be all kinds of materials and equipment on its inner surface. But the greatest proportion of the lunar mass is concentrated in the central part of the sphere, in its core . . .

Vasin and Shcherbakov suggest that at some point in the far distant past, for reasons unknown, a highly advanced alien civilisation captured a small world and hollowed out its interior, transferring their entire civilisation to this new habitat and setting off into the unknown reaches of the Galaxy. At some point, this world-ship was guided (whether by living or automatic pilots) into orbit around a young planet that would one day be called by its inhabitants 'Earth'.

One could speculate at length on the reasons for this grand

endeavour. Perhaps the aliens' homeworld had suffered some kind of cataclysm which made it impossible for them to stay there. Perhaps they were combatants in an ancient interstellar war between supercivilisations, and found themselves on the losing side.

Perhaps the impetus for that great voyage was not disaster but the desire to explore and gather knowledge of the Universe. Science fiction writers have often written about 'generation ships', colossal vessels travelling slower than the speed of light and carrying tens of thousands or even millions of people, most of them living out their entire lives between worlds.

It is easy to imagine scenario upon scenario; but do any of them bear the slightest relation to reality? It is up to the reader to decide whether the 'Spaceship Moon' theory makes any sense, or if it really belongs in the realms of science fiction. If it is true, however, then the Moon may hold wonders beyond imagining: the artefacts and records of an entire civilisation, awaiting the intrepid human explorers of the (hopefully) near future.

Would that we could choose which mysteries were real!

18

FROM THE EDGE OF REALITY

MYSTERIOUS CREATURES OF THIS WORLD AND OTHERS

Centuries ago, when the world was not nearly as well travelled and explored as it is today, cartographers inscribed a phrase on their maps to denote places where men of their nations had yet to set foot. That phrase was: Here Be Dragons. It is a strange phrase, at once elegant and sinister, hinting at mysteries and dangers scarcely imaginable lurking in places unseen by 'civilised' eyes.

It is a strange irony that in the modern world – a world crisscrossed by roads and railways, strewn with cities, overflown by airliners full of tourists and watched by the satellites of several nations – there are still places where the phrase 'Here Be Dragons' might well apply. Some of these places are in vast tracts of seldom-travelled wilderness, regions which have yet to know the taming hand of human civilisation, while others are to be found – strangely, incredibly – in the very midst of our cities.

Against all logic and common sense, it seems that the 'dragons' still exist, and are to be encountered by the unwary and hapless in every corner of our world. The light of reason and technology should have banished such things long ago, relegating them to an ignorant and superstitious past when ghosts, werewolves and vampires were feared, when fairy spirits visited lonely cottages in the dead of night bringing strange gifts or stranger misfortunes; when seafarers spoke in hushed tones of the vast krakens lurking in dark ocean deeps.

Those times and their stories should have gone for ever: tales to frighten children, and no more. And yet strange creatures have been encountered by countless witnesses across the globe, in city and wilderness alike. What are we to make of such reports? Are they nothing more than the stuff of overactive imaginations? Are they lies told to garner fame by those who do not deserve it? Are they simply misidentifications of ordinary animals seen in extraordinary or unfamiliar conditions? Each of these explanations might suffice, particularly in the minds of sceptics who maintain that such things do not and cannot exist.

However, some reports are so bizarre that one finds it difficult to imagine how otherwise sane and rational people could have made them. Surely they must have been aware of the disbelief and ridicule with which their reports would be met. As we shall see in the following pages, it may be too easy to resort to disbelief and ridicule in the face of these strange tales. It may be better to ask: can they be true? And if so, what should be done about them?

MOTHMAN

The word 'monster', of course, has many usages. Anything large, wild and deadly can be called a monster, whether it be a shark, an enraged grizzly bear, a shapeless jellyfish trailing vast, lethal tentacles . . . the list of dangerous animals which would be best not encountered by fragile humans is long indeed. How much more frightening when the monster is shaped like a man, but is most certainly not a man? How much more sickening the dread when that which is encountered should not exist, even in nightmares?

One of the most frightening and perplexing of monstrous humanoids is undoubtedly the creature that came to be known as the 'Mothman'. Those who were intrigued by the recent film *The Mothman Prophecies* starring Richard Gere would do well to read the book of the same name by the veteran paranormal researcher John Keel, on which it is based. Enjoyable as the film is, it did not (and perhaps could not) do justice to the sheer weirdness of the original events.

It began on a cold November night in 1966 in the small, sleepy town of Point Pleasant, West Virginia. Two young couples, Mr and Mrs Roger Scarberry and Mr and Mrs Steve

Mallette, were driving through an ammunitions dump left over from the Second World War, a place known as the 'TNT Area', which lay about seven miles outside of town, and which was a popular spot for young lovers anxious to be out from under the watchful gaze of their parents. As their car passed the vast, ramshackle edifice of an abandoned power plant, the four young people caught a glimpse of something that shouldn't have been there, something shaped like a man but somewhat larger, with great wings that appeared to be folded against its back.

The driver, Roger Scarberry, slowed the car as they passed, and he and his passengers gazed in disbelief at the apparition standing by the roadside. The thing then turned away from the road and walked, with a strange shuffling gait, towards the entrance to the power plant. They looked at each other, and each could see the fear and confusion on the others' faces. Without waiting to be told, Roger floored the accelerator and the car took off with a screech of tyres away from the TNT Area.

They thought they had left the bizarre and frightening 'thing' behind, but that was far from the case, for they quickly realised that the winged creature was following them. Looking up through the car's windows, they saw its man-like shape flying through the air above them, keeping pace with their car even though, in his panic, Roger was nudging 100 mph along the deserted road. Even more bizarrely, the creature did not appear to be flapping its wings.

When they finally (and thankfully) reached the city limits of Point Pleasant, the creature abandoned its chase and disappeared into the night. They drove straight to the Mason County Sheriff's Office and breathlessly told Deputy Millard Halstead what had happened. Halstead later said: 'I've known them all their lives. They've never been in any trouble. I took them seriously.' The Deputy decided to take a look around the TNT Area and the power plant; but he returned saying that everything seemed to be in order – and there were certainly no flying monsters to be seen.

The tale might have ended there, and been consigned to the anonymous, musty archives of the paranormal, had not the witnesses insisted on holding a press conference at which they described their experience. It was picked up by the wire services, and, for want of anything else to call it, a newspaperman dubbed the creature 'Mothman', after the popular TV show *Batman*. Although the creature bore precious little resemblance to a moth, the name stuck, capturing the public imagination and bringing the town of Point Pleasant to national prominence.

A few days after that first sighting, Mrs Marcella Bennett was driving through the TNT Area with Mr and Mrs Raymond Wamsley, on their way to see their friends the Thomas family. When they arrived at the Thomas house (one of the few homes lying within the area), Marcella got out of the car, carrying her two-year-old daughter in her arms. As she approached the house, she caught a movement out of the corner of her eye, and turned to see something rising up from the ground, as if it had been lying down and had been roused by their presence.

Marcella later described it as tall and grey in colour, with terrible, glowing red eyes. In abject horror, she and the Wamsleys rushed for the house and desperately hammered upon the front door. Ralph and Virginia Thomas were not at home, but their children, Rickie, Connie and Vickie, were. Startled and fearful, the children let the visitors in, and Raymond Wamsley called the police as the creature shuffled onto the porch and peered in through the windows. By the time the police arrived, the creature had departed.

Following this encounter, more and more people began to see the Mothman, which was usually described as being between 5 and 7 feet tall, with a wingspan of about 10 feet. Most reported that it did not appear to have a head, and that its huge, glowing red eyes were in its shoulders. Bizarre as this sounds, Mothman was not the strangest interloper to haunt the normally quiet and peaceful community of Point Pleasant. There were numerous UFO sightings in the area throughout Mothman's reign of terror, which has led many researchers to suggest that there may have been some connection – although precisely what that connection might be is open to conjecture. In addition, there were several encounters with 'Men in Black', those weird personages who have frequently been reported to harass and threaten witnesses to UFO-related events.

In the past, sightings of strange creatures have been considered omens of imminent disaster, and, tragically, Point Pleasant was no different. On 15 December 1967, just over a year after the first sighting of Mothman, the Silver Bridge, an old suspension bridge spanning the Ohio River near Point Pleasant, collapsed during the rush hour, killing nearly 50 people.

BATWOMAN

Flying humanoids have been seen all over the world throughout history. One of the most curious modern cases occurred near Da Nang, Vietnam, in July 1969. The main witness was an American soldier named Earl Morrison, who was serving at the time with the US 1st Division Marine Corps. Morrison was on guard duty with two other men one night, when at about 1.30 a.m. they saw an object coming towards them through the dark sky. According to Morrison:

> We saw what looked like wings, like a bat's, only it was gigantic compared to what a regular bat would be. After it got close enough so we could see what it was, it looked like a woman. A naked woman. She was black. Her skin was black, her body was black, the wings were black, everything was black. But it glowed. It glowed in the night – kind of a greenish cast to it . . . She started going over us, and we still didn't hear anything. She was right above us, and when she got over the top of our heads, she was maybe six or seven feet up . . . we watched her go straight over the top of us, and she still didn't make any noise flapping her wings. She blotted out the moon once – that's how close she was to us. And dark – looked like pitch black then, but we could still define her because she just glowed. Real bright like. And she started going past us straight toward our encampment. As we watched her – she had got about 10 feet or so away from us – we started hearing her wings flap. And it sounded, you know, like regular wings flapping. And she just started flying off and we watched her for quite a while.

Morrison was questioned by the researcher Don Worley, who wrote up the account in the British journal *Flying Saucer Review* in June 1972. The witness believed that the creature was covered with short fur, and that the hair on her head was black and straight. In addition, 'the skin of her wings looked like it was moulded onto her hands', and her wings rippled in a way which suggested that she didn't have any bones in her arms.

In their book *Alien Animals*, Janet and Colin Bord note that the only other winged female figure they have in their records is the Gwrach-y-rhibyn, a banshee-like entity from Welsh folklore. They go on to quote the description given by the Welsh folklorist Marie Trevelyan, who describes the Gwrach-y-rhibyn as:

> . . . having long black hair, black eyes, and a swarthy countenance. Sometimes one of her eyes is grey and the other black. Both are deeply sunken and piercing. Her back was crooked, her figure was very thin and spare, and her pigeon-breasted bust was concealed by a sombre scarf. Her trailing robes were black. She was sometimes seen with long flapping wings that fell heavily at her sides, and occasionally she went flying low down along watercourses, or around hoary mansions. Frequently the flapping of her leathern bat-like wings could be heard against the window-panes.

THE CORNISH OWLMAN

Strange and terrifying creatures are not only to be found in distant and exotic countries. In 1976 the English county of Cornwall was besieged not only by a bizarre flying humanoid which came to be called 'Owlman', but also by the sea monster known as Morgawr.

In the small south-coast village of Mawnan, two sisters, June and Vicky Melling, saw what they described as a 'big feathered birdman' hovering over the church bell tower. A couple of months later, two girls named Sally Chapman and Barbara Perry were camping in the woods near the church when they heard a strange noise and saw a figure standing a few yards away. According to Sally:

> It was like a big owl with pointed ears, as big as a man. The eyes were red and glowing. At first, I thought it was someone dressed up, playing a joke,

trying to scare us. I laughed at it, we both did, then it went up in the air and we both screamed. When it went up, you could see its feet were like pincers.

Were these sightings just children's fantasies? If so, they were contagious, for the very next day another girl, Jane Greenwood, who was on holiday in the area, also saw the Owlman. Jane wrote of her experience in a letter she sent to the local newspaper, the *Falmouth Packet*.

It was Sunday morning and the place was in the trees near Mawnan Church, above the rocky beach. It was in the trees standing like a full-grown man, but the legs bent backwards like a bird's. It saw us and quickly jumped up and rose straight up through the trees.

My sister and I saw it very clearly before it rose up. It has red slanting eyes and a very large mouth. The feathers are silvery grey and so are his body and legs. The feet are like big, black crab's claws.

We were frightened at the time. It was so strange, like something in a horror film. After the thing went up there were crackling sounds in the tree tops for ages.

Later that day we spoke to some people at the camp-site, who said they had seen the Morgawr Monster on Saturday, when they were swimming with face masks and snorkels in the river, below where we saw the bird man. They saw it underwater, and said it was enormous and shaped like a lizard.

Our mother thinks we made it all up just because we read about these things, but that is not true. We really saw the bird man, though it could have been somebody playing a trick in very good costume and make-up.

But how could it rise up like that? If we imagined it, then we both imagined the same thing at the same time.

THE GLOBSTERS

Strange and frightening as they may be, creatures such as Mothman and Owlman at least possess a recognisable humanoid shape. Even more bizarre and mystifying are the so-called 'Globsters' (the name was coined by the pioneer monster researcher Ivan T. Sanderson), which have been dredged up from the ocean depths, and occasionally discovered washed up on beaches, and which can only be described as shapeless lumps of flesh and skin.

The 'Globster of Margate' was described by another monster-hunter, Bernard Heuvelmans, who described how, on 1 November 1922, a farmer named Hugh Ballance noticed a disturbance in the sea just offshore near the South African town of Margate. Peering into the distance, Ballance saw an apparent battle between two whales and what he described as a 'sea monster' looking like a 'polar bear'. Ballance later said: 'This creature I observed to rear out of the water fully 20 feet and to strike repeatedly with what I took to be its tail at the two whales, but with seemingly no effect.'

The battle continued, and before long a fair-sized crowd had gathered on the shore to watch the strange spectacle. Eventually, the whales departed, leaving their enemy floating apparently lifeless on the ocean's surface. That night, the sea brought it ashore, and the next morning the crowds returned to find a creature the likes of which none of them had ever seen.

It was 47 feet long, 10 feet wide and 5 feet high, with a 10-foot-long tail at one end, and at the other, a strange appendage reminiscent of an elephant's trunk. The trunk was 14 inches in diameter and 5 feet long, and its tip looked like a pig's snout. If this were not startling enough, the onlookers realised that the creature was covered with an 8-inch-long coat of white hair. It hardly need be said that there is no known marine animal with a 'trunk', and yet this appendage has been reported on several other Globsters found elsewhere.

In their book *Unexplained Phenomena*, Bob Rickard and John Michell quote the great American anomalist Charles Fort, who collected many tales of unknown creatures being washed up by the sea. 'It may be,' wrote Fort, 'that there have been several finds of remains of a long-snouted animal that is unknown to the paleontologists, because, though it has occasionally appeared here, it has never been indigenous to this earth.'

This is an intriguing idea – if rather difficult to prove! For his part, Ivan Sanderson also thought the hypothesis worth pursuing, and suggested that some UFOs might actually be living creatures, perhaps indigenous to space itself.

Sanderson also investigated another Globster case, which occurred on Tasmania's west coast in August 1960. Following a powerful storm, a large mass of flesh was washed up on the beach at Temma, where it was discovered by a cattleman named Ben Fenton. The mass was roughly circular and measured some 20 feet across; there was a raised central mound, and the body also contained large slits which looked a little like gills.

Eventually, word of the creature reached the city of Hobart and the Commonwealth Scientific and Industrial Research Organisation (CSIRO). One of its employees, Bruce Mollinson, decided to travel to the beach at Temma to examine the thing for himself. He later wrote an account of his expedition, in which he stated: 'It wasn't fish, fowl or fruit. It wasn't whale, seal, sea elephant or squid. Sunfish, some had guessed, but this creature did not have fins. Devil ray? A devil ray has a mouth and teeth.'

Mollinson finally decided to risk a theory: it was, he suggested, a totally unknown animal that may have come from the deep undersea caverns off the coast of Tasmania. Journalists from far and wide offered their own opinions. Perhaps it was a prehistoric creature that had been 'thawed out', a real-life analogue of any number of Hollywood B-movies. Perhaps it was the remains of a mammoth (the thing had a thick hairy pelt) that had been released from its icy Antarctic prison and been washed ashore. Perhaps it really was a space monster.

Goaded into action by Mollinson's report, the CSIRO mounted its own expedition. In *Unexplained Phenomena*, Rickard and Michell note that the organisation's conclusions, published in the Hobart *Mercury* of 19 May 1962, seem to refer to a completely different carcass to the one Mollinson examined. For one thing, it was much smaller, and the CSIRO concluded that it was nothing more bizarre than a lump of decomposing whale meat.

Mollinson, who was not present during this investigation, had his doubts that they had examined the same object he had.

Rickard and Michell suggest that 'the CSIRO team, which had omitted to take along any of the original discoverers, could have stumbled across a rotting whale and mistaken it for the body in question'.

For their part, the people who had first seen the carcass were quite adamant that it was not a whale, not even a small part of a whale. It was like nothing they had ever seen.

THE ELECTRIC HORROR

At first sight, the stories associated with the elegant townhouse at 50 Berkeley Square, London, might appear to describe an 'ordinary' haunting (if there can be such a thing). It is only when one delves a little deeper into the house's history that one finds hints of something altogether more mysterious and terrifying – something, truly, from the edge of reality.

During the nineteenth century, the house was considered to be the most haunted in the whole of London, and there were many strange and eerie stories associated with it. No one was quite sure exactly who or what inhabited the house, since it was said that all those who encountered it either died immediately or else were driven hopelessly insane.

When someone wrote to the magazine *Notes & Queries* in November 1881, enquiring as to whether the house really was haunted, another correspondent replied with reference to an article in the *Mayfair* magazine of the previous May:

> The mystery of Berkeley Square still remains a mystery. The story of the haunted house in Mayfair can be recapitulated in a few words; the house contains at least one room of which the atmosphere is supernaturally fatal to body and mind ...
>
> The very party-walls of the house, when touched, are found saturated with electric horror. It is uninhabited save by an elderly man and woman who act as caretakers; but even these have no access to the room. This is kept locked, the key being in the hands of a mysterious and seemingly nameless person, who comes to the house every six months, locks up the elderly couple in the basement, and then unlocks the room, and occupies himself in it for hours.

The house at 50 Berkeley Square had possessed an evil reputation for many years previous to the *Mayfair* article. Perhaps its most famous victim (at least in the annals of the supernatural) was the 20-year-old dandy Sir Robert Warboys, who was drinking with some friends in a tavern in Holborn one night in 1840, when the conversation turned to the ill-rumoured house. Sir Robert dismissed the tales as childish nonsense; his friends, however, were not so sure, and suggested that he put his money where his mouth was, as it were, and spend a night in the place.

Sir Robert gladly accepted the challenge, and that very night he visited the house's landlord, John Benson, to tell him of his plan. Benson, who lived on the ground floor, tried to talk him out of staying in the 'haunted room' on the first floor; but Sir Robert would hear no objections, and demanded to be put up there for the night. Reluctantly, the landlord agreed, but only on the condition that he take a loaded pistol with him, and that if he should see anything unusual, he would pull on a cord connected to a bell in the landlord's quarters.

Benson handed Sir Robert a pistol, and bade him goodnight. The young man then sat down in the haunted room and happily waited for something to happen.

Shortly after midnight, the landlord was woken up by the frantic ringing of the bell, followed by a single gunshot. Tumbling from his bed and racing upstairs, he burst into the room to find Sir Robert crouched in a corner. He was dead, apparently from shock, his eyes bulging and his lips drawn back from his teeth in a rictus of terror. Kneeling beside him, Benson looked in the direction of the dead man's gaze, and saw a single bullet hole in the opposite wall. Clearly, in that last awful moment of his life, Sir Robert had fired at *something*, but just what that 'something' was, no one could say.

Three years later, the house claimed two more victims. Edward Blunden and Robert Martin were sailors from Portsmouth, who were wandering around the streets drunkenly after a night on the town. They had spent all their money on drink, and had none left for lodgings, so when they saw the 'To Let' sign outside the house in Berkeley Square, they thought that fate had smiled on them. After all, why should they sleep on the streets when there was a perfectly serviceable house standing empty for them?

So the sailors broke into the house and, finding the ground floor a little too damp for their liking, made their unsteady way upstairs to the first floor . . . and the haunted room.

No sooner had they fallen asleep than they were awoken by the sound of shuffling on the staircase; then, to their horror and disbelief a shadowy figure entered the room. Martin fled screaming from the room and the house. He found a policeman walking past in the square outside and told him what was happening. At that moment, there was a sound of shattering glass, and the body of Edward Blunden plunged from the upstairs window, impaling itself on the iron railings in front of the house.

The ongoing correspondence in *Notes & Queries* resulted in more tales coming to light. In one case, a well-to-do family had hired the house for the London season, as their daughters were to be 'brought out'. One of the girls was engaged, and her fiancé was invited to stay at the house. The family instructed one of their maids to prepare a room for him (the reader may be able to guess which room it was).

The maid was still attending to her task at midnight on the evening before the fiancé's arrival, at which time a loud, piercing shriek was heard coming from the room. Roused from their sleep, the family rushed into the room where they found the maid crouched upon the floor in convulsions, staring madly at one corner. She was taken to St George's Hospital, where she died the next morning after gibbering repeatedly of something 'horrible' in the room.

The fiancé arrived the following day as planned. After telling him the whole dreadful story, the family agreed that it would not be a good idea for him to sleep in that room. However, he shrugged the matter off and insisted on sleeping there. Reluctantly, the family agreed, but insisted in their turn that he sit up until after midnight, and to ring if he needed assistance for any reason.

At midnight, the family heard the bell ring once, but only faintly. A few minutes later, however, it rang again, this time loudly and continuously. Rushing into the room, they found their young guest crouched upon the floor in exactly the same spot as the unfortunate maid, and gazing wide-eyed at the same far corner of the room. The fiancé survived the experience with his sanity apparently intact; but he would never speak of what he had seen in that room: it was, he said, simply too horrible to describe.

As mentioned, a correspondent to *Notes & Queries* wrote of the party-walls of the house being 'saturated with electric horror'. This description seems to have resulted from another letter to the magazine which told of a ball held at Number 49 next door in the summer of 1880. A lady and her partner were sitting against the wall separating the house from Number 50, when she suddenly stood up and whirled around to face the party-wall. Her partner was about to ask her what was wrong, when he too jumped away from the wall. Both were deeply unsettled: they had both felt suddenly terribly cold while sitting against the wall, and both had had the distinct impression of something unspeakable watching them from behind.

There are other tales told of the house, which are not so much ghost stories as tales of cosmic horror, such as are found in the works of writers like H.P. Lovecraft and Clark Ashton Smith. One such tale tells of some nameless thing, too horrible to describe, slipping up and down the stairs and leaving a foul-smelling trail behind it, like some monstrous and unearthly slug.

Whatever the ill-rumoured history of 50 Berkeley Square, whatever the horrors said to have walked there, today it seems that the house is safe. It is currently occupied by Maggs Bros Ltd, the antiquarian booksellers, who have never experienced anything out of the ordinary there. Today, the haunted room is next to the accounts department, and the house's fearsome supernatural history seems to have given way to the tediousness of day-to-day commerce.

SECRET AGENTS FROM THE EDGE OF REALITY

There are many weird and wonderful creatures associated with the UFO phenomenon, all of which are apparently non-human. Well . . . perhaps not quite all, for there is a class of entity which has been encountered on many occasions by UFO witnesses, and which seems to be not quite human and not quite alien. These are the so-called 'Men in Black', or MIBs.

On the evening of 11 September 1976, Dr Herbert Hopkins was alone in the family house in Orchard Beach, Maine, his wife and son having gone to the cinema. Dr Hopkins had been investigating an alleged UFO encounter experienced by a young man named David Stephens through the use of regressive hypnosis. He received a telephone call from a man claiming to be a member of a UFO organisation in New Jersey, who asked if he might discuss the Stephens case with Hopkins. The doctor was a little surprised that the man knew about the case; nevertheless, he agreed and invited the man to the house.

Hardly had Hopkins put the phone down, when the caller appeared at the front door. He was the most singular individual Hopkins had ever seen: completely bald, he was dressed in an immaculate black suit, black tie and white shirt. He had no eyebrows or eyelashes, and when he later rubbed his mouth, it became evident that he was wearing lipstick. The man entered the house, walking stiffly and awkwardly, and when invited to take a seat in the living room, he sat perfectly still.

In a flat, expressionless voice, the visitor began to ask Hopkins various questions about the Stephens case; and then something truly extraordinary happened. The man said that Hopkins had some coins in his pocket, and asked him to take one out and hold it in the palm of his hand. Somewhat nonplussed, Hopkins did so. As he watched in amazement, the coin's colour changed from bright silver to light blue, and it began to grow blurred. It then gradually faded away and vanished.

Suspecting some kind of clever trick, Hopkins said he was impressed, and asked the man to make the coin reappear. Chillingly, in his flat, toneless voice, the man replied: 'Neither you nor anyone else on this plane will ever see that coin again.' (Note the intriguing use of the word 'plane' rather than 'planet'.)

The man then began talking about the famous Barney and Betty Hill abduction case, which had occurred in New Hampshire in 1961 (see Chapter 24). He asked Hopkins how Barney Hill had died, and Hopkins replied that as far as he knew Barney had died of a stroke. 'No,' said the man. 'Barney died because he had no heart, just as you no longer have your coin.'

Having made this bizarre and disturbing statement, the man told Hopkins to destroy all the materials relating to his investigation of the Stephens case.

The encounter took yet another weird turn when the visitor said, with apparent difficulty: 'My energy is running low . . . must go now . . . goodbye.' He then left, walking unsteadily around the corner of the building. As soon as the man was out of sight, Hopkins saw a blue-white light shining on the driveway. He ran around the corner, but of the mysterious Man in Black there was no sign.

When Hopkins' wife and son returned from the cinema, they found him in a state of near panic, sitting at the kitchen table with a gun, and with all of the lights in the house switched on. Hopkins later complied with the MIB's instructions, and destroyed all of his materials on the Stephens case.

This is just one case of many, each as bizarre as the others. Who, or what, are the Men in Black? After examining the many encounters with them, researchers have built up a picture of them, based on the testimony of witnesses. Frequently they are described as being quite short – about 5 feet 6 inches or less – with olive complexions and strangely slanted eyes. They are always dressed the same way, in conservative black suits with white shirts, and frequently claim to be government agents.

Their mode of transport is not always as exotic as that employed by Dr Hopkins' visitor: usually they are described as arriving in immaculate luxury cars (Cadillacs in the United States, Bentleys or Jaguars in the United Kingdom) which nevertheless are several years out of date.

When a Man in Black comes to call, the witness frequently reports feeling ill at ease, and yet incapable of turning the visitor away. It is as if the percipient has been placed in a light trance, in which he or she is unable to act as they normally would.

Men in Black frequently act in an extremely bizarre (even silly) fashion. In one case, an MIB asked his host for some jelly, and then attempted to drink it from the bowl, while his female companion (whose lipstick had been inexpertly applied, as if she were unfamiliar with the concept of make-up) looked at the spoon she had been given as if she had never seen such a thing before.

MIBs usually visit people who have had UFO or other paranormal encounters, and threats such as that made against Dr Hopkins are also common, with the MIBs demanding that their hosts stop talking about their experiences. Strangely, however, they never seem to make good on those threats, which are frequently made in corny, B-movie terms. For instance, one MIB told his host to stop talking about the UFO he had seen, if he wanted his wife 'to stay as pretty as she is'.

According to the veteran paranormal researcher Brad Steiger:

> After witnesses of UFO or paranormal phenomena have received a visit from the MIB and surrendered whatever evidence the alleged agents of the government demanded, the harassment is by no means over. Telephones ring at all hours with threatening or nonsensical mechanical voices. Television and radio programs are interrupted by weird signals, claiming to be of alien origin. Network video and audio are blotted out to be replaced by images of robed, sometimes cowled, figures, who instruct the witnesses to continue to cooperate and to keep all UFO information confidential. In exchange for this silence and cooperation, the mysterious entities sometimes promise the witnesses key roles in marvellous projects that will benefit all humankind.

Although some researchers have suggested that MIBs may be aliens disguised (albeit rather ineptly) as humans, they frequently seem to have more in common with ghosts than extraterrestrials. Their appearance and behaviour suggest that they are somehow dislocated in time. While some MIBs are entirely normal-looking, and may well be human beings following their own obscure agendas, the more bizarre specimens could conceivably be psychic manifestations of the paranoia that gripped America in the 1950s and '60s. It could be that the more outlandish MIBs represent a psychic 'echo' of that past national trauma, just as certain houses seem to retain a kind of 'recording' of violent or tragic events.

19

INTO OBLIVION

MYSTERIOUS VANISHINGS

It was to have been an evening of celebration and joy. During Christmas Eve, 1909, five inches of snow had fallen, covering the ground in an undisturbed carpet of white as far as the eye could see. At the Thomas farm near the town of Rhayader, Wales, a number of friends had gathered for an evening of carol singing. As the evening wore on and midnight approached, the Thomas family and their guests roasted chestnuts on the glowing embers of the fire and prepared to celebrate the arrival of Christmas Day. Around the hearth of Owen Thomas were his family, several close friends, the minister and his wife, the local veterinarian and an auctioneer from one of the neighbouring villages.

At about 11 o'clock, Owen Thomas noticed that the water bucket was empty. In those days, rural houses relied on wells for their drinking water. Mindful of his guests, whose joyful singing would doubtless bring on a thirst, Thomas turned to his 11-year-old son, Oliver, and asked him to take the bucket out to the well and fill it.

Oliver was a good lad, and despite the cheery warmth of the house, he was happy to pull on his boots and overcoat, and venture out into the icy night air. He had been gone only a few seconds, when the carol singing was interrupted by his frantic cries for help.

Shocked by the unexpected sounds, everyone rushed out of the house and into the snow-covered yard, which Oliver had been crossing on his way to the well only a few moments before.

The minister had seized a paraffin lantern, which he now held up, casting a flickering light on the empty yard. Oliver had gone. His footsteps, which could be clearly seen in the five-inch-deep snow, extending in a straight line in the direction of the well, stopped suddenly 75 feet away, as if something had swooped down from the sky and scooped up the unfortunate lad.

In abject horror, the Thomas family and their guests listened as Oliver screamed: 'They've got me! Help, help! They've got me!'

Everyone tried to place the direction from which the terrified cries were coming, and soon they realised that Oliver's voice was coming from *above* them. The minister raised the lantern as high as he could, attempting to cast some light into the pitch-black sky. It was no use. Oliver was nowhere to be seen, but his cries continued to drift down to them, tormenting his horrified parents as they gradually grew fainter and fainter. Eventually, they ceased altogether, leaving the Thomas family and their guests alone in the yard beneath the silent, frozen sky.

There were no other tracks in the snow, and no signs of a struggle of any kind. The water bucket Oliver had been carrying lay about 15 feet away from where his footprints ended.

The following day, police from Rhayader appealed to local residents for help in locating the boy, and a search party was formed, which examined the Thomases' well and then scoured the surrounding countryside. The Thomas family and their guests of the previous evening were closely questioned, but the only conclusion to which the police could come was that Oliver Thomas had been seized on his way to the well, and taken directly upwards.

Although this conclusion was clearly not very satisfactory to the police, the physical evidence pointed rather strongly to it; but who, or what, could have done such a thing? No known animal could have snatched the boy without leaving tracks; certainly no known bird could have carried off an 11-year-old. Neither could any known flying machine have done so in 1909.

The only clue – and it is a tenuous one – is what Oliver Thomas himself cried out as he was being taken up into the air: 'They've got me!' Not 'it', not 'something', but 'they'. As the American paranormal researcher Brad Steiger states in his description of the case, this implies that Oliver got a look at whoever had taken him. Oliver was never seen again. If he knew who or what his kidnappers were, he took the knowledge with him, into the sky . . .

There is an almost identical story from across the Atlantic, regarding the strange disappearance of a 16-year-old boy named Charles Ashmore. On a November evening in 1878, Charles walked out of his family's farmhouse near Quincy, Illinois, to fill a bucket with water from a nearby spring. When he did not return, his father, Christian Ashmore, took a lantern and went to find him. Charles' footprints were clearly visible in the new-fallen snow, but came to an abrupt end 75 yards from the house. Ashmore looked around the yard in confusion, and then went to the spring. The water was covered with a thin layer of unbroken ice.

The Ashmore family were grief-stricken by the sudden, unexplained disappearance of their son; but their horror and torment were increased when, four days later, Charles's mother went to the spring for water and, at the place where his footprints in the snow had ended, heard his voice calling out to her. In fear and confusion, she wandered around the yard, trying in vain to ascertain the direction from which the voice was coming. When she was questioned later, she maintained that the voice had definitely been that of her son, although she could not make out what he was saying.

For several months afterwards, members of the Ashmore family heard Charles's voice every few days. The voice was quite distinct, and yet it seemed to come from a great distance, so that its words could not be made out. Eventually, the voice grew fainter, and by the mid-summer of 1879, Charles Ashmore's voice was heard no more.

Once again, there are considerable similarities between this story and that of the strange disappearance of a Tennessee farmer named David Lang.

Lang's farm lay a few miles outside the town of Gallatin, and he lived there with his wife, Emma, and their two young children, George and Sarah. On the afternoon of 23 September 1880, David Lang left the farmhouse and started across the 40-acre pasture to check on his horses. Emma watched him from the porch where George and Sarah were playing happily with a wooden horse and wagon which their father had brought back from Nashville that morning.

At that moment, a horse and buggy came into view on the lane leading up to the house; the family could see their friend, Judge August Peck, at the reins, and George and Sarah immediately stopped playing and started to jump up and down with excitement, since Judge Peck always brought them presents whenever he visited. As he caught sight of the horse and buggy, Lang stopped, waved to his friend and turned around to return to the house.

And then, in full view of his wife, children and friend, David Lang vanished into thin air.

Emma screamed, and the children, startled and confused, stood and watched mutely as their mother rushed to the spot where Lang had been. Peck jumped down from his buggy and raced across the pasture, arriving at the spot within moments of Emma. They found no sign of David Lang. There were no trees or bushes in which to hide, no hole down which he might have fallen, not a single clue to indicate what might have happened to him. He was gone, as if he had never existed.

Peck ran to the side yard and raised the alarm by ringing a large bell that stood there, and while Emma was led into the house, still screaming hysterically, the neighbours scoured the pasture for any sign of Lang. News of the bizarre disaster spread rapidly, and by nightfall scores of people had come to offer their assistance in the search. They searched every square foot of the pasture, stamping their feet on the dry, hard ground in the hope of uncovering a hole into which he might have fallen – but they found nothing.

In the weeks that followed, Emma Lang was bedridden from the shock of what she had witnessed. The county surveyor examined the pasture where Lang had vanished, and found it to be supported throughout by a thick layer of limestone: there were no caves or sinkholes into which the unfortunate farmer might have fallen. Emma Lang never fully recovered, and refused to allow a funeral or even a memorial service to be held for her husband, whom she believed was still alive and would someday return.

Eventually, she allowed Judge Peck to rent out the farm, with the exception of the 40-acre pasture in front, where her husband had met his unimaginable fate.

That, however, was not the end of the story. Several months

later, in April of 1881, Lang's two young children were in the pasture, when they noticed a circle of stunted yellow grass at the precise spot where their father had vanished. That evening, Sarah Lang returned to the spot alone, and called out to her father. To her astonishment, she heard David Lang's voice calling in return, calling out to her for help, over and over again, until it finally faded away, never to be heard again.

Equally strange is the tale of James Worson, a shoemaker who lived in Leamington, England. Although honest and well liked, Worson was fond of making foolish bets when he had had a little too much to drink. It was one of these bets which seems to have cost Worson his very existence.

On 3 September 1873, Worson was boasting to some friends of his prowess as an athlete, and bet them a sovereign that he could run all the way to Coventry and back, a distance of some 40 miles. He set out at once, closely followed in a horse-drawn cart by the three friends with whom he had made the wager.

His occasional bragging was not entirely misplaced: Worson was indeed a fit and strong man, and the first few miles of his run passed without incident. Occasionally his three friends in the cart would offer him a few good-natured words of encouragement.

Suddenly, Worson seemed to stumble, then fell headlong forward and, with a terrible shriek, vanished into thin air. He did not fall to the earth: he vanished before touching it, and no trace of him was ever discovered.

Frightened and bewildered, Worson's three friends returned to Leamington and informed the authorities of what had happened. Needless to say, they were immediately taken into custody; however, since Worson's body was never found, and his friends were known to be of good standing and had always been considered truthful, they were set free, and the mystery of James Worson's disappearance remains unsolved to this day.

LIFE IMITATING ART: THE STRANGE FATE OF AMBROSE BIERCE

These are intriguing tales to be sure – and not a little frightening, if one pauses to consider that the strange calamities which befell Owen Thomas, Charles Ashmore, David Lang and James

Worson might happen to anyone at any time – even to you who are reading these words at this very moment. That is to say, they *would* be intriguing and frightening – if there were a scrap of truth to them, which there isn't. Notwithstanding the fact that they have been told many times in many books, magazine articles and Internet websites devoted to the paranormal, these particular tales of supernatural vanishings are, unfortunately (or perhaps fortunately), not true.

Rather, they were inspired by one of the most curious, idiosyncratic and popular of ghost- and horror-story writers, Ambrose Bierce – and some are verbatim transcriptions of his stories. Bierce's life was as strange as any of his fictions, and his death – or disappearance – was equally mysterious; indeed, it has become one of the most celebrated and puzzling of unexplained vanishings, and is a fitting inclusion in our brief survey of this fascinating subject.

Ambrose Gwinett Bierce was born on 21 June 1842 in Horse Creek, Meigs County, Ohio, the tenth of thirteen children. His parents were deeply religious, and his father, an unsuccessful farmer, had a profound love of literature which, along with religion and extreme poverty, was the chief influence on Bierce's early life. He quickly came to despise his parents for their poverty and religious beliefs, and this is doubtless a contributing factor to the deep bitterness and cantankerousness which characterised his personality. His sense of humour was so black as to be virtually ultraviolet, as evidenced in the opening sentence of his short story 'An Imperfect Conflagration': 'Early in 1872 I murdered my father – an act that made a deep impression on me at the time.'

Bierce's mordancy, bitterness and extremely low opinion of humankind were reinforced by his experiences in the American Civil War, in which he fought on the Union side. According to the writer David Stuart Davies: 'The horrible accidents and cruel twists of fate that he encountered during the conflict stimulated his imagination into creating what was regarded as a new genre – the short story that begins as a war story but then subtly changes into a psychological tale of terror.'

He was a successful soldier, achieving the rank of major, but his response to an offer of a large sum in back pay when the war ended was typical; he refused it, saying: 'When I hired out as an assassin for my country, that wasn't part of the contract.' He went to San Francisco in 1867 and, following a brief stint as a night-watchman at the Mint, decided on a career in journalism and managed to secure a job on the *San Francisco News Letter*. From then on, Bierce's rise in the city's literary circles was swift and spectacular (he numbered Mark Twain and Bret Harte among his friends), and his position in San Franciscan society was cemented by his marriage in 1871 to the socialite Mary Ellen Day.

With the $10,000 his father-in-law gave to him as a wedding present, Bierce took Mary to England, where they lived for three years, spending some time in the town of Leamington, where the story of James Worson, 'An Unfinished Race', is set. Returning to America in 1875, they again took up residence in San Francisco, where Bierce maintained his celebrity as a writer and journalist, editing and contributing to a number of newspapers and journals.

During this period, while contributing his famous 'Prattle' column to William Randolph Hearst's newspaper the *San Francisco Examiner*, Bierce wrote his most famous book, *The Devil's Dictionary*, a pessimistic, cynical and hilarious collection of definitions, including the following:

- ADMIRATION: our polite recognition of another man's resemblance to ourselves.
- BORE: a person who talks when you wish him to listen.
- CYNIC: a blackguard whose faulty vision sees things as they are, not as they ought to be.
- EGOTIST: a person of low taste, more interested in himself than in me.
- IDIOT: a member of a large and powerful tribe whose influence in human affairs has always been dominant and controlling.
- JUSTICE: a commodity which in a more or less adulterated condition the State sells to the citizen as a reward for his allegiance, taxes and personal services.
- LOVE: a temporary insanity curable by marriage.

The turn of the new century brought personal disasters for Bierce; his son Leigh died of pneumonia brought on by alcoholism and his wife, Mary, divorced him on the grounds of abandonment. Although he continued writing, Bierce became increasingly cantankerous and restless, engaging less and less with the world around him. In 1912, after revisiting the battlefields of the Civil War, he decided to continue south into Mexico, where another civil war was raging. If his final letter from Chihuahua in December 1913 is anything to go by, his thoughts were of suicide: 'If you hear of my being stood up against a Mexican wall and shot to rags, please know that I think it's a pretty good way to depart this life. It beats old age, disease, or falling down the cellar stairs. To be a Gringo in Mexico – ah, that is euthanasia.'

No one knows what happened to Ambrose Bierce, although some have suggested that his friendship with the bandit-turned-revolutionary Pancho Villa turned sour (probably because of a well-aimed Bierce insult), resulting in his being executed. One of his friends maintained that he had put a gun to his head on the edge of the Grand Canyon, pulled the trigger and fallen into its depths ('an appropriate tomb for his gigantic ego'), while a more macabre and less likely story is that he was captured by a Mexican tribe and boiled alive, his remains being worshipped – an idea that would doubtless have appealed to his dark sense of humour.

What really happened to Ambrose Bierce is likely to remain a mystery, however; and, as David Stuart Davies notes: 'In many ways it is fitting for a man who created so many dark and surprising scenarios in his fiction to end his life with a question mark – his own personal, autobiographical twist in the tail.'

It is an undeniable fact that every year, hundreds of thousands of people go missing across the globe. Many return unharmed, having decided temporarily to retreat from the world for their own reasons; many are later proved to have met with accidents or foul play; many are abducted and murdered; and some simply vanish without a trace for all time. Many vanishings have been cited as proof of alien interlopers, and even portals leading to other worlds and other dimensions through which human beings sometimes inadvertently stumble. Is there any truth to these bizarre theories?

A STRANGE VORTEX

In December 1873, a young well-to-do couple were arrested for disorderly conduct at Bristol railway station. The man, Thomas Cumpston, had earlier discharged a pistol at the hotel in which he and his wife Annie were staying, frightening the other guests. In fact, after staggering into the station dressed in their nightclothes, the couple had demanded that the superintendent call the police, for they claimed to have escaped from 'a den of rogues and thieves'. The police were duly called, and had little choice but to take Thomas and Annie Cumpston into custody until the truth of the evening's events could be ascertained.

The story they had to tell was so bizarre and extraordinary that it was later reported in *The Times*, although the paper's editor felt obliged to present it under the title 'Extraordinary Hallucination'. The Cumpstons' testimony was as follows.

On the evening of Monday 8 December 1873, the couple were travelling from Clifton to Weston-super-Mare, and booked into the Victoria Hotel, which was situated near the railway station, so that they might get an early start and continue their journey. The Cumpstons were from Leeds, and were a well-off, respectable couple not easily given to bizarre flights of fancy. Perhaps a little oddly, Thomas habitually carried three knives and a pistol; however, as Rodney Davies notes in his book *Supernatural Disappearances*, this 'may be explained by the fact that handguns could then be bought over the counter at gun shops, and because the city streets of Victorian Britain were hazardous places in which to be'.

The Cumpstons went to bed at about midnight. They had been asleep only for an hour or so when they were woken up by voices apparently coming from the room next door. Irritated, they roused the landlady, but the voices had ceased by the time she entered the room, and she suggested that they must have been mistaken. There seemed to be nothing for it but to return to bed, which the Cumpstons did. There were no more disturbances for a couple of hours or so, but then several events occurred which caused the young couple to flee the hotel in terror.

According to the *Bristol Daily Post*, which reported the affair:

About three or four o'clock they heard worse noises, but what they were they had no idea. The floor seemed to be giving way, and the bed also seemed to open. They heard voices, and what they said was repeated after them. Her husband wished her to get out of the way. The floor certainly seemed to open, and her husband fell down some distance, and she tried to get him up. She asked him to discharge his pistol to frighten anybody who might be near, and he fired his revolver into the ceiling. They got out of the window, but she did not know how, being so frightened; and when they got to the ground she asked him to fire off another shot, which he did.

This rather prosaic account hardly does justice to what must have been a truly terrifying event. According to the Cumpstons, the floor of their hotel room collapsed or opened into a whirlpool-like hole, filled with impenetrable darkness, from which strange, mocking voices issued, and which repeated everything they said (although one is tempted to wonder whether these were really the couple's own voices, echoing distortedly from inside the strange vortex). Thomas Cumpston was pulled half into the swirling hole, and had to be dragged out of it by his wife.

As the vortex sucked in various items in the room, the Cumpstons realised that they could not reach the door without risking a plunge into its unimaginable depths, and so they opted for the window – even though it was a 12-foot drop to the street outside. Risking broken limbs, they jumped from the window and ran in blind panic to the nearby railway station where they raised the alarm.

The Cumpstons were arrested for disorderly conduct, and appeared in court the following day, where they told their astonishing story. They sent a telegraph to a friend in Gloucester, asking him to come to Bristol to vouch for their good characters. The court accepted his statement, and the Cumpstons were discharged without a fine.

This is a curious story indeed, but is it true? As we have already seen, reports of strange disappearances are often spurious, and involve people who do not seem to have existed in the first place! While researching *Supernatural Disappearances*,

Rodney Davies went to Bristol Central Library, where one of the librarians, Elizabeth Shaw, told him that the Victoria Hotel stood at 140 Thomas Street, 'although its name was changed to the Bute Arms in 1876'. Davies also established that Thomas and Annie Cumpston were a real couple, who lived at 35 Virginia Road, Leeds, according to the 1881 census.

THE VANISHING VILLAGE

Many researchers into apparently supernatural disappearances have wondered whether they might be connected with another great paranormal mystery (perhaps the greatest of the twentieth century): that of UFOs. Alleged alien abductions have entered the public imagination, and the 'Grey' alien has become a powerful cultural icon as instantly recognisable as any world-famous trademark. Could some bizarre vanishings be attributable to the activities of sinister alien forces operating in our world?

One of the strangest and most spine-chilling of possible abduction cases occurred (so the story goes) on the shores of Lake Anjikuni, Canada, in the winter of 1930. The case is spine-chilling because it involved the disappearance of an entire Eskimo village of 1,200 souls.

It began with the sighting by a fur-trapper named Armand Laurent of a strange, irregularly shaped object tumbling end-over-end through the sky. Laurent and his two sons were outside their cabin when the object drifted overhead. Later, they described it as continuously changing shape as it flew, appearing first as a cylinder, then as a bullet-like projectile.

A few days after this sighting, two Royal Canadian Mounted Police (RCMP) officers arrived at the Laurents' cabin, and asked if they might shelter there for the night. They were, they said, on their way up to Lake Anjikuni, where there was a 'kind of problem'. When Laurent mentioned the strange object he and his sons had seen tumbling through the sky, the Mounties were interested, and asked which direction the object was moving in. Laurent replied that it was headed towards Lake Anjikuni.

For Armand Laurent, that was the end of the affair, and the memory of the strange thing whirling through the cold sky gradually faded.

Joe Labelle was not so lucky. Joe was also a trapper, and a few days previously he had been approaching the village on the shore of the lake, looking forward to spending some time with his many friends there. However, the pleasant anticipation he felt as he made his way across the frozen landscape gradually turned to creeping apprehensiveness as he drew nearer.

Stopping at the edge of the village, he shouted a greeting; but there was no one to be seen in the normally bustling community. Labelle found this hard to understand. Peering through the icy gloom towards the lakeshore, he could make out a few kayaks lying on the beach, battered and half torn to shreds by the wind and the waves.

Labelle entered the village and moved slowly and carefully among the huts, glancing with increasing dread at the empty doorways, with their caribou-skin flaps billowing forlornly in the frigid wind. For more than an hour he wandered from hut to hut, searching in vain for signs of the village's inhabitants. In some huts he discovered pots of caribou stew hanging cold and congealed over the ashes of long-dead fires; in others he found sealskin garments, some belonging to children, one of which had an ivory needle hanging from it. A mother must have been mending it, Labelle surmised, when something happened to make her suddenly stop ...

To Labelle, however, the most disturbing thing was the presence of the Eskimos' rifles, standing abandoned by the doors of the huts. To a seasoned trapper, this was truly incomprehensible: an Eskimo's rifle is his single most important possession. Not one of the villagers would have gone out into the vast wilderness beyond the village without it.

And yet, gone they had, without their only means of protection.

Labelle couldn't stand to spend another moment in that place made strange and horrible by its sudden, unexplained desertion. He fled across the frozen landscape to the nearest telegraph office and sent a report to the RCMP base in Churchill. The Mounties who were dispatched to the lake persuaded Labelle to go with them. Reluctantly, he agreed. He did not want to go back, but he *did* want to find out what had happened to the Eskimos, many of whom were his friends.

About 100 yards from the edge of the village, they found the Eskimos' sled dogs. They had been tied to some stunted trees, and were frozen solid, having died of starvation.

Mystery was compounded by mystery, and yet the strangest and most frightening aspect of the puzzle was about to reveal itself. On the far side of the village, the Eskimos had buried a deceased member of their community under a traditional cairn of stones. When Labelle and the Mounties came to the grave, they saw that it was empty. The stones had been placed to one side in two neat piles, and the ground had been opened and the body removed. None of the investigators could understand why this had happened – or indeed *how*, since the ground was frozen solid and hard as iron.

A major investigation was launched, and a nationwide search conducted with the aid of experienced trackers, but no trace of the villagers was ever found. The only conclusion that seemed reachable was that at some point everyone had left their homes, apparently attracted by something that was happening outside. They had left their rifles, their dogs, their food and clothing, and had not returned. Their kayaks had been left behind also, meaning that they had not set out across the lake for some unknown destination.

The American author and abductee Whitley Strieber wove the Lake Anjikuni incident into the narrative of his 1989 novel *Majestic*, and came up with an intriguing possibility regarding the mystery of the empty grave. Since the Eskimos had great respect for their dead, might not the deceased man have been taken away also, as a sign of respect for their beliefs? This, of course, raises the question: who or what took them away?

Although Eskimos do migrate suddenly from their settlements, they do not do so without the means to ensure their survival. The questions surrounding this strange episode have never been answered, and the RCMP file on the disappearance of the Lake Anjikuni community remains open to this day.

THE LAST FLIGHT OF FREDERICK VALENTICH

Many UFO researchers have seen a sinister and significant connection between Armand Laurent's sighting of an apparent UFO and the unexplained disappearance of the Eskimos who had made their home on the shores of Lake Anjikuni. In the case of Frederick Valentich, the connection between a UFO and a subsequent disappearance is even more powerful.

On 21 October 1978, 20-year-old Valentich took off from

Moorabbin Airport in Victoria, Australia, in his single-engine Cessna 182 aircraft. His destination was King Island, about one hour's flight away. At about 7 p.m., Valentich watched the approach of an unidentified aircraft, which hovered above him, causing his Cessna's engine to falter. He radioed air traffic controller Steve Robey at Melbourne Air Flight Service, informing him of what was happening. The transcript (here slightly abridged) makes disturbing reading:

VALENTICH: Is there any known traffic below five thousand [feet]?

MELBORNE FLIGHT SERVICE: No known traffic.

V: . . . seems to be a large aircraft below five thousand.

MFS: What type of aircraft is it?

V: I cannot confirm. It has four bright, it seems to me like landing lights . . . the aircraft has just passed over me at least a thousand feet above.

MFS: Roger, and it is a large aircraft? Confirm.

V: Er . . . unknown due to the speed it's travelling. Is there any Air Force aircraft in the vicinity?

MFS: No known aircraft in the vicinity.

V: It's approaching now from due east towards me . . . It seems to me that he's playing some sort of game. He's flying over me two to three times . . . at speeds I can't identify.

MFS: Roger. What is your actual level?

V: My level is four and a half thousand. Four five zero zero.

MFS: Confirm that you cannot identify the aircraft.

V: Affirmative.

MFS: Roger. Stand by.

V: It's not an aircraft. It is – [brief silence].

MFS: Can you describe the, er, aircraft?

V: As it's flying past, it's a long shape . . . cannot identify more than . . . before me right now, Melbourne.

MFS: And how large would the, er, object be?

V: It seems like it's stationary. What I'm doing right now is orbiting, and the thing is just orbiting on top of me. Also, it's got a green light and sort of metallic. It's shiny on the outside . . . It's just vanished . . . Would you know what type of aircraft I've got? Is it a military aircraft?

MFS: Confirm that the, er, aircraft just vanished.

V: Say again.

MFS: Is the aircraft still with you?

V: . . . Approaching from the southwest . . . The engine is rough idling. I've got it set at twenty-three twenty-four, and the thing is [coughing].

MFS: Roger. What are your intentions?

V: My intentions are, ah, to go to King Island. Ah, Melbourne, that strange aircraft is hovering on top of me again . . . It is hovering, and it's not an aircraft.

Seventeen seconds of silence followed this exchange; then there was a loud metallic scraping sound . . . and nothing more. Frederick Valentich was neither seen nor heard from again. The conclusion of the Aircraft Accident Investigation Summary, released in May 1982, was as follows:

Location of occurrence: Not known.

Time: Not known.

Degree of injury: Presumed fatal.

Opinion as to cause: The reason for the disappearance of the aircraft has not been determined.

When the UFO researcher, or ufologist, Bill Chalker began to investigate the case, and asked for clarification of the Bureau of Air Safety's findings, the first Assistant Secretary (Air Safety Investigation) G.V. Hughes wrote to him:

A great deal of consideration has been given to what Mr Valentich might have been looking at when he described his observations. A considerable number of suggestions have been put forward by persons inside and outside this Department. All have been examined. The department is not aware of any other official body having undertaken such an investigation into this occurrence . . .

As you correctly state . . . the RAAF is responsible for the investigation of reports concerning 'UFO' sightings, and liaison was established with the RAAF on these aspects of the investigation. The decision as to whether or not the 'UFO' report is to be investigated rests with the RAAF and not with this Department.

Chalker requested and was given access to the RAAF UFO files in Canberra. According to the historian of ufology, Jerome Clark:

He examined what was represented as every UFO-related document the Directorate of Air Force Intelligence (DAFI) had in its possession. He found nothing on the Valentich case. The DAFI Intelligence Liaison Officer told him that the RAAF had not investigated the incident because the Department of Aviation had not asked it to do so. The RAAF professed to see it as an 'air accident/air safety' matter, the officer said; he then expressed his private view that Valentich had crashed after he became disoriented.

In 1982, an independent film producer named Ron Cameron was approached by two divers who claimed to have located the wreckage of Valentich's Cessna on the seabed off Cape Otway. They showed him some photographs of a plane with the Cessna's registration marks, and offered to sell them to him, along with the plane's exact location, for $10,000. They also claimed that Valentich's body was not in the plane.

Cameron replied that he would need further verification before he handed over that much money. Jerome Clark writes:

As Cameron contemplated a salvage operation, he heard from the Department of Aviation (DoA), which stated it had to be involved in such activity; after all, the aircraft was still the subject of an open air-accident investigation. It added, however, that it wanted to keep a low profile. A meeting was arranged, but in the wake of growing publicity, the DoA shied away, fearing a media circus. The divers also backed away from Cameron, after complaining that statements he had made in a radio broadcast indicated he had doubts about their honesty. Cameron would claim he had assured the divers that he had implied no such thing, but soon the divers and Cameron were out of communication. Later, when he tried to contact them, he could not find them.

The Valentich disappearance remains unsolved, and there is no evidence that he faked his own death. It has been suggested that his plane got into trouble, ditched in the sea and sank immediately. However, the Cessna 182 is designed to float on impact with water. In addition, as ufologist Timothy Good states: 'VHF radio would not be able to transmit below 1,000 feet from the aircraft's position of ninety miles from Melbourne, and Valentich's communications with the Flight Service Unit were loud and clear to the last word, as was the seventeen-second burst of "metallic" noise which followed.'

THE VANISHING LIGHTHOUSE KEEPERS

The Flannan Isles lie about 30 kilometres to the west of the Outer Hebrides, in the midst of the wild grey waters of the north Atlantic. The main group of seven islands, known also as the Seven Hunters, have always been considered strange and dangerous places, and many believed that they were home to some mysterious and malignant spirit or force, frequently referred to as the 'Phantom of the Seven Hunters'. They also presented a more mundane threat to shipping in the area, and this led to the construction of a lighthouse on the largest island of the group, Eilean Mór, in 1899.

Eilean Mór rises from the roiling waters like a great, clenched fist, and the lighthouse sits atop its humped northern section, flashing its warning signal twice every 30 seconds from an elevation of some 100 metres above sea level. Although it is the largest of the Flannans, it is still very small, approximately 800 metres long by 500 wide. It is here that an event occurred which has entered the realms of maritime legend, and still stands as one of the strangest and eeriest of mysteries.

At Christmas 1900, the lighthouse on Eilean Mór was manned by three keepers: James Ducat, the head lighthouse keeper, or Principal; Thomas Marshall, the 2nd Assistant; and Donald McArthur, a part-time (or Occasional) keeper, who was doing duty for William Ross, the 1st Assistant, who was on sick leave. Keeping watch in lighthouses was a lonely occupation. During the day, lightkeepers' duties included cleaning the station, and painting if necessary. They also had to maintain the equipment, making sure that the light was in proper working order, the oil fountains and canteens filled, the lenses polished and the wicks properly trimmed. At night, the keepers were required to keep watch in the lightroom to make sure that the light was working properly and flashing correctly to character, and also to keep a constant fog watch, and be ready to operate the fog signal in the event of poor visibility. The lightkeepers had to be men of many talents and abilities, with a good working knowledge of engines, and also a healthy respect for the capricious and destructive power of the sea. They had to be good handymen and decent cooks, as well as genial and pleasant companions – a more important requirement than one might think, since they were required to man rock stations for four weeks at a time.

It was on the afternoon of 26 December 1900 that the Northern Lighthouse Board (NLB) lighthouse tender *Hesperus* approached Eilean Mór. Aboard was the relief keeper Joseph Moore, by all accounts a sensitive and conscientious young man, who would be doing a month's duty on the island. As the tender approached, the crew expected to see the lighthouse flag flying as a sign of welcome. But no flag flew above the station. Assuming that Ducat, Marshall and McArthur had not seen them coming, the ship's captain, James Harvie, sounded the fog horn several times, but still there was no reply, and no one came out to greet them. Harvie ordered a rocket to be fired. Still nothing.

The sea was high and rough, and eventually the *Hesperus* managed to lower a rowing boat, which tied up at one of the island's two landings. Joseph Moore went in the boat, along with two members of the tender's crew, Archie Lamont and William McCormack. Moore was the first to go ashore, while his fellows minded the boat.

The young relief keeper hurried up the stone steps that had been cut into the island's flank, and arrived at the entrance gate to the small compound containing the light and the adjoining dwelling. He found the gate closed, as was the door leading to the house's kitchen. Moore entered and looked around the kitchen, which was clean, the dishes having been washed and put away. He glanced at the fire, and noted that it contained nothing but cold ashes. Everything seemed normal, except that the clocks had stopped, and one of the chairs was turned over and lying on the floor.

Moore called out to the men, but there was no answer. Thinking that they must all have fallen sick and taken to their beds, he checked the bedrooms, calling out the keepers' names as he did so. But there was no one there, and the beds were all neatly made. The whole place was steeped in a strange silence made worse by the howling of the wind outside.

Realising that something serious had happened, Moore rushed back to the landing and alerted Lamont and McCormack. Other members of the *Hesperus*' crew joined in a search of the island, but no trace of the three keepers was found. On the East Landing, where the rowing boat had tied up, everything was in order, but the West Landing presented a different picture. A box used for holding mooring ropes and tackle had gone, and some of the ropes it had contained lay strewn upon the rocks. There was also a handrail, which wound up alongside the stone steps leading to the top of the island, which was attached to the rock by sturdy stanchions; but a long section of the handrail appeared to have been ripped out and twisted strangely.

Lighthouse keepers are required to maintain log entries, which are made first in chalk on a slate, and then transferred to the logbook later. The last entry at the lighthouse on Eilean Mór was made on the slate by James Ducat on 15 December, which implies that he and the others disappeared later on that day, before they had a chance to light the lamps.

Two of the three sets of oilskins and sea boots were missing, which suggested that two of the men had gone outside.

But why did the third man neglect to put on his own gear? Moreover, it was one of the cardinal rules of lightkeeping that no lighthouse should ever be left unmanned, and thus it was almost inconceivable that all three keepers would be outside at the same time – unless the last man was called to attend some dreadful emergency.

An investigation was conducted by Superintendent Robert Muirhead of the Northern Lighthouse Board. His conclusion was that, since there was a powerful storm on the night of 14 December, James Ducat would have been worried about the landing ropes and tackle. Six months earlier, the tackle on the West Landing had been carried off by a fierce storm, and had had to be replaced. Ducat would have known that if it happened again, it would make relief all the more difficult.

Late in the afternoon, the wind began to drop, and so Ducat and one other man put on their oilskins and sea boots, and made their way to the west side of the island, leaving the third man in the lighthouse, according to the rules. When they came to the place where the tackle was stored, one of the men (we have no way of knowing who) was seized by a sudden violent wave and carried into the sea. The other man ran back to the lighthouse, shouting through the kitchen doorway what had happened. The third man sprang to his feet, overturning his chair, and rushed out without taking his own oilskin.

The two men returned to the West Landing, and tried to rescue their comrade from the sea, but another wave crashed upon them and carried them off to their deaths.

In part, Superintendent Muirhead's report reads:

> After a careful examination of the place, the railings, ropes, etc. and weighing all the evidence which I could secure, I am of the opinion that the most likely explanation of the disappearance of the men is that they had all gone down on the afternoon of Saturday 15 December to the proximity of the West landing, to secure the box with the mooring ropes, etc. and that an unexpectedly large roller had come up on the Island, and a large body of water going up higher than where they were and coming down upon them had swept them away with resistless force.

I have considered and discussed the possibility of the men being blown by the wind, but, as the wind was westerly, I am of the opinion, notwithstanding its great force, that the more probable explanation is that they have been washed away as, had the wind caught them, it would, from its direction, have blown them up the Island and I feel certain that they would have managed to throw themselves down before they had reached the summit or brow of the Island.

On conclusion of my enquiry on Saturday afternoon, I returned to Breasclete [Shore Station], wired the result of my investigations to the Secretary and called on the widows of James Ducat, the Principal Keeper, and Donald McArthur, the Occasional Keeper.

I may state that, as Moore was naturally very much upset by the unfortunate occurrence, and appeared very nervous, I left A. Lamont, seaman, on the Island to go to the lightroom and keep Moore company when on watch for a week or two.

If this nervousness does not leave Moore, he will require to be transferred, but I am reluctant to recommend this, as I would desire to have at least one man who knows the work of the Station . . .

In conclusion, I would desire to record my deep regret at such a disaster occurring to Keepers in this Service. I knew Ducat and Marshall intimately, and McArthur the Occasional, well. They were selected, on my recommendation, for the lighting of such an important Station as Flannan Islands, and as it is always my endeavour to secure the best men possible for the establishment of a Station, as the success and contentment of a Station depends largely on the Keepers present at its installation, this of itself is an indication that the Board has lost two of its most efficient Keepers and a competent Occasional.

I was with the Keepers for more than a month during the summer of 1899, when everyone worked hard to secure the early lighting of the Station before winter, and, working along with them, I appreciated the manner in which they performed their work. I visited Flannan Islands when the relief was made so lately as 7th December, and have the melancholy recollection that I was the last person to shake hands with them and bid them adieu.

Is this really what happened to the lighthouse keepers of Eilean Mór? Over the years, many researchers have puzzled over the mystery, and have come up with some quite bizarre theories, including that the men were carried off by pirates or sea serpents, that they were abducted by aliens, or fell through a rift between this reality and another. Some have also suggested that the legendary force known as the Phantom of the Seven Hunters took them to an unknown fate as retribution for humanity's trespassing into its domain.

In spite of the logic and rationality of Superintendent Muirhead's conclusion as to the tragic fate of Ducat, Marshall and McArthur, the true reason for their disappearance will probably never be known.

THE BENNINGTON TRIANGLE

To all those who were frightened silly by the film *The Blair Witch Project*, it may come as an unpleasant surprise to learn that there is a region in the north-eastern United States with a similarly evil reputation, a place where strange lights have been seen in the night skies, and people have vanished without a trace, never to be seen again. That place is the wilderness around Glastenbury Mountain and the town of Bennington in rural Vermont; it has been nicknamed the Bennington Triangle by local researcher Joseph Citro, who has spent many years collecting folktales, legends and strange reports from the deep woods in the region.

Glastenbury Mountain has always had a sinister reputation, going right back to the days of the Native Americans, who feared the place and shunned it. They described it as a cursed place, where all four winds met, and went there only to bury their dead.

When the first European settlers arrived, they too sensed something horrible and otherworldly about the place; they spoke of strange lights in the skies, weird noises in the dead of night and unidentifiable odours drifting through the woods. Aside from the noises and lights and odours, plain bad luck seemed to have made its home there: the original town of Glastenbury fell victim to disease, bad weather and mysterious deaths, and was eventually 'unorganised' in 1937.

Glastenbury seemed to be a place where bad things happened as a matter of course. In 1892 Henry MacDowell murdered a fellow mill worker named Jim Crowley in a drunken brawl. MacDowell was declared insane and incarcerated in the Waterbury Asylum, from which he promptly escaped and vanished into the hills, never to be seen again. However, it would be another half century or so before the area began to acquire the broader notoriety it possesses today.

The first disappearance occurred on 12 November 1945. Seventy-four-year-old Middie Rivers led four men on a hunting trip into the mountains. Rivers was a veteran of the wild woodlands; he had spent many years hunting and fishing there, and knew the region as well as a city dweller knows his own street. That afternoon, as they made their way back to their camp, Rivers pulled ahead of the others, heading towards Long Trail Road . . . and vanished. Police and volunteers searched the area extensively, but no trace of the experienced guide was ever found – except for a single bullet which was found beside a stream. It was concluded that Rivers had stooped to take a drink, and that the bullet had slipped out of his pocket. Logical enough, but it shed precious little light on what had happened to Middie Rivers.

Just over one year later, on 1 December 1946, 18-year-old Paula Welden set out from her dormitory at Bennington College. She had decided to go for a walk on the Long Trail – and walked into oblivion. Many people saw her leave, including Ernest Whitman, who worked for the *Bennington Banner* newspaper, and an elderly couple who were also out walking on the Long Trail, and who later confirmed that they had seen the young woman walking about 100 yards ahead of them. They added that Paula Welden walked around a bend in the trail, and when they reached the bend, she had disappeared.

A massive search was begun, with more volunteers scouring the area for the missing girl. The FBI also became involved, and a $5,000 reward was put up for information leading to her safe return, all to no avail. Paula Welden was never seen again, and rumours began to circulate that she had gone to Canada with a lover, or perhaps was living as a 'wild woman' in the hills (just why she would do this, nobody knew).

Perhaps the strangest disappearance associated with the Bennington Triangle is that of James Tetford, a resident of Bennington Soldiers' Home, who was returning by bus to Bennington after visiting relatives in St Albans. Witnesses saw him get on the bus, and later the other 14 passengers confirmed that he was asleep in his seat when it pulled into the stop before Bennington. But when the bus pulled into Bennington itself, he was no longer in his seat; his belongings were still in the luggage rack, but of James Tetford there was no sign. The date of his disappearance was 1 December 1949, exactly three years after Paula Welden went missing. He was never seen again.

On 12 October 1950, eight-year-old Paul Jepson became the next victim of the dark forces apparently at work in the area. His parents were caretakers of a local dump, and his mother was tending to the family's pigs, leaving Paul unattended for no more than an hour. When she went to check up on him, he had gone, never to be seen again. The boy's father later stated that the lad had expressed a strange fascination with the nearby mountains. As with the other victims, an intense search was made of the surrounding country, using bloodhounds. The dogs eventually tracked his scent to a certain stretch of road, and then lost it. This suggested that Paul may have been picked up by someone driving along the road; however, the point where the bloodhounds lost his scent was the exact location of Paula Welden's disappearance four years previously.

The next disappearance occurred just over two weeks later. Freida Langer was hiking with her cousin, Herbert Eisner, when she stumbled into a stream. She decided to run back to their camp to change clothes. However, when Herbert returned to the camp, Freida was not there; nor had anyone else at the camp seen her arrive. It was broad daylight, and Freida knew the area well, having hiked there on many previous occasions. Once again, a search was mounted, both on foot and by plane and helicopter.

However, on 12 May 1951, Freida's badly decomposed body was found in the area of the Somerset Reservoir – an area that had been extensively searched only a few months earlier. She was the only victim of the Bennington Triangle to be found.

After 1950, the disappearances in the region stopped as mysteriously as they had begun, and no one has gone missing in the Bennington Triangle since then. As might be expected, many theories have been put forward to explain the strange events. Perhaps the most bizarre and fanciful is the idea that there is some kind of 'dimensional horizon' in the area, into which hapless people occasionally stumble, never to return (apart from Freida Langer, that is). Other researchers have noted the local Native American legend of an enchanted stone which instantly consumes anyone who stands on it. Aliens have also been blamed, of course, as has the 'Bennington Monster', a Bigfoot-like creature which some claim to have spotted moving through the deep, dark woods.

A rather more prosaic theory is that a serial killer may have been active in the region, although serial killers tend to target a certain type of individual, and the people who disappeared in the Bennington Triangle were both male and female, and were of diverse ages.

It would seem that the Bennington Triangle will keep its secrets, perhaps for all time, with those unfortunates who have disappeared near Glastenbury Mountain merely added to the ever-lengthening roster of people who seem, for no rational reason, to have vanished into oblivion.

20

AN EXPERIMENT WITH TIME

THE MONTAUK PROJECT

U FO conspiracy theories are, by their very nature, weird and outrageous – none more so than the splendidly bizarre conspiracy known as 'the Montauk Project'. Based near the popular resort town of Montauk at the easternmost end of Long Island, New York, this ultra-secret project apparently succeeded not only in opening time portals to different periods in the past and future, but in transferring people and equipment to other planets and dimensions.

These outlandish claims are presented in a series of books by Preston Nichols and Peter Moon, which began with *The Montauk Project* in 1992. It must be said that information on the Montauk Project cannot claim the soundest of provenances, having its roots in the legendary Philadelphia Experiment of October 1943. Although it's an intriguing story, and would make for pretty good science fiction, the Philadelphia Experiment is now accepted by many (but by no means all) paranormal researchers as spurious. Nevertheless, it has become firmly entrenched in the mythology of paranormal conspiracy theories, and since it is so crucial to the genesis of the Montauk Project, it is worth a brief review for the sake of background.

AN EXPERIMENT IN INVISIBILITY

What came to be known by UFO researchers as the Philadelphia Experiment took place aboard the USS *Eldridge*, which was berthed in the Philadelphia Navy Yard. The initial intention was to use powerful electromagnetic fields in an attempt to render the ship invisible to radar; however, the results of the field manipulations were totally unexpected. Not only did the USS *Eldridge* become invisible to radar, it also became invisible to *sight*. And literal invisibility was not the only achievement, for the ship was reported to have suddenly appeared in the Navy yards at Norfolk, Virginia: it seemed that the US Navy had unwittingly succeeded in teleporting the USS *Eldridge* a distance of 200 miles.

For the crewmen on board the ship, the results of the experiment were disastrous and appalling: when the ship returned to its berth in the Philadelphia Navy Yard, many of them had become fused with the decks and bulkheads, while those who survived had become hopelessly insane and babbled about encountering strange creatures while in the 'hyperspace' between Philadelphia and Norfolk.

This chilling information was first revealed to the astronomer and pioneering UFO researcher Morris K. Jessup by a man who called himself Carlos Allende or Carl Allen. In July 1955, a copy of Jessup's book *The Case for the UFO* was anonymously mailed to the Chief of the Office of Naval Research (ONR) in Washington, DC. The book had been heavily annotated in three different colours of ink, apparently by the same person. The Special Projects Officer of the ONR, Commander George Hoover, was so intrigued by the information contained in these annotations, which included details of the origin and propulsion of the UFOs, that he called Jessup to Washington to discuss them.

Jessup was shocked by the annotations to his book because of their similarity to the information with which he had been supplied by Carlos Allende, information that referred to the invisibility experiments conducted aboard the USS *Eldridge*. Subsequently, the ONR commissioned the Varo Manufacturing Company (based in Texas and involved in high-tech military research) to print a limited edition of the annotated text of *The Case for the UFO*.

THE MONTAUK PROJECT

According to Preston Nichols and Peter Moon, the Montauk Project was the culmination of 40 years of experiments that followed the alleged teleportation of the USS *Eldridge* and the destruction, both mental and physical, of her crew. Nichols claims that these experiments included 'electronic mind surveillance and the control of distinct populations', and culminated in 1983 with the opening of a doorway through time to the Philadelphia Experiment in 1943.

The project was conducted at the Air Force base on the grounds of Fort Hero, which had been officially decommissioned in 1969, but was subsequently reactivated 'without the sanction of the US Government'. In the early 1970s, Preston Nichols, an electronics engineer and resident of Long Island, had been experimenting with parapsychology, and was working with a group of psychics. Having come to the conclusion that telepathy operates on a principle similar to that of radio waves, he was intrigued to discover that all of the psychics in the group suddenly and inexplicably lost their ability at the same time every day.

Nichols suspected that this was the result of some form of electronic interference, and set about searching for its nature and source. He discovered that the psychics' abilities were being impeded by a transmission on the 410–420 MHz wavelength, which seemed to originate in the officially abandoned Montauk Air Force Base. When he visited the base, he found it to be active and tightly guarded, and he couldn't get anywhere near the large radar antenna that was broadcasting the disruptive transmissions.

In 1984 Nichols was informed by a friend that the base had again been abandoned, so he went back with one of his psychic colleagues and found large amounts of equipment strewn around, as if the personnel had left in a hurry. They also came upon a man living rough in the abandoned buildings who had apparently been a technician involved in the Montauk Project. He told them that there had been an important experiment the previous year that had resulted in the appearance of a strange, violent 'beast' which had frightened away all the personnel and forced them to abandon the programme. Nichols was astonished when the man added that he recognised him as his boss on the project, for Nichols certainly had no memory of working there.

Following his visit to the base, Nichols was further unnerved when a stranger arrived at his house and insisted that Nichols had been the Assistant Director of the Montauk Project. Nichols began to suspect that he had been living in an alternate 'timeline', of which he had been consciously unaware.

Acting on a hunch, Nichols went to the basement of the building in which he worked (he was employed by a defence contractor on Long Island), which contained a high-security area. He had been experiencing strange, unsettling feelings while in certain parts of the building, and was anxious to find out why. To his surprise, when he walked up to the security guard and handed him his low-level pass, the guard wordlessly exchanged it for a high-level pass that would enable him to enter the restricted area. Once inside, he discovered an office which provoked an unpleasant visceral response, and entered. On the desk was a name plate with his name on it, and the title 'Assistant Project Director'.

Later, Nichols constructed what he calls a 'Delta Time Antenna' on the roof of his house, which brought an explosion of memories from the alternate timeline in which he had been working on the Montauk Project. With the help of these rediscovered memories, he was able to establish that the main thrust of the project had been to manipulate time in order to open up portals to distant places and times, using technology based on designs provided by aliens from the Sirius star system. This was achieved by means of a psychic, who sat in what was known as the 'Montauk Chair'. The chair was hooked up to a battery of advanced computers which decoded his thoughts and projected them to the desired location. The psychic would then open up a tunnel through spacetime, through which people could walk.

The controllers of the Montauk Project began to explore time, opening portals to the past and future. The tunnel, or 'vortex', had a helical shape, like a corkscrew which, Nichols said, 'twisted and took turns until you'd come out the other end'. The people sent through the vortex were instructed to observe their new surroundings and then return to Montauk. According to Nichols, 'winos or derelicts' were frequently grabbed off the streets, sobered up for a week or so, and then sent into the vortex. The reason for this was that if there was a power failure while a vortex was open, the person inside would be lost for ever, cast adrift in time and space with no hope of returning. Since this was a not-infrequent occurrence, the 'explorers' had to be people who would not be missed if anything went wrong.

Many of these victims were adolescent boys, who were snatched from the streets of big cities like New York. Nichols claims that they were all tall, blond and blue-eyed, conforming to the Aryan stereotype (Nichols briefly alludes to a neo-Nazi connection at Montauk). These boys were invariably sent forward through time to AD 6037. They would arrive in a ruined city, at the centre of which was a square containing a golden statue of a horse. Each traveller was required to read the hieroglyphic inscriptions on the statue, and report any impressions he got from them. Nichols suspects that there was some form of technology in the statue, which the Montauk controllers wanted to investigate. At any rate, the inscriptions on the statue appeared to be different each time a group was sent through.

As if all of these claims were not outrageous enough, Nichols adds that attempts were also made to reach the interior of the alien city which allegedly lies ruined in the Cydonia region of Mars. Apparently, an ultra-secret human colony had already been established on Mars to investigate the ruins, but had been unable to find a way into the colossal buildings, so the Montauk scientists decided to open a portal directly into the underground region beneath one of the largest structures, the so-called 'DiPietro Molenaar Pyramid'. An effort was also made to search the past for the builders of the Martian monuments. Apparently, they were encountered in about 125,000 BC, although Nichols says he has been unable to discover anything else about this particular experiment.

The expedition into the DiPietro Molenaar Pyramid resulted in the discovery of technology that came to be known as 'the Solar System Defence', for which the pyramid acted as a kind of antenna. This technology was found to be interfering with the research being conducted on Mars, so the scientists at Montauk had it shut off 'retroactive to 1943'. Nichols points out that it was at about this time that UFO sightings began to increase, starting with the so-called 'foo fighters' that plagued both Allied and German aircraft in the closing years of the Second World War.

Nichols says that he and several other scientists on the Montauk Project were becoming increasingly apprehensive about the irresponsible nature of the spacetime experiments. For this reason, they decided to go ahead with a contingency plan to sabotage the entire project. A psychic sitting in the Montauk Chair was given a secret signal, whereupon he released a terrifying monster from his subconscious (reminiscent of the 'monster from the id' in the science fiction classic *Forbidden Planet*). The monster was transmitted into three-dimensional reality, and proceeded to smash parts of the base to pieces. The controllers attempted to banish the entity from normal spacetime by completely shutting down all the power generators on the base.

This was apparently the end of the Montauk Project. The base was decommissioned in 1983; all the personnel, including Nichols, were debriefed and brainwashed to forget everything that had happened.

As with many other conspiracy scenarios, there is a powerful element of occultism at work here. The idea of alien involvement is combined with parapsychology and the paranormal, so that the boundaries between them are obscured. To the disinterested observer, this is both fascinating and irritating, because a large number of additional elements are brought into play which, at first glance, ought to have nothing to do with 'science', however outlandish and spurious. In the Montauk Project, the basic scenario – experiments in teleportation and the manipulation of time – mixes uneasily with the idea of alien-supplied technology, operated with the aid of psychics and their paranormal gifts.

There is an additional factor in all this which needs to be mentioned, since it further illustrates the fusion of apparently contradictory ideas. In *Montauk Revisited*, their sequel to *The Montauk Project*, Nichols and Moon describe how the experiments at Montauk had a powerful antecedent in the form of the so-called 'Babalon Working', a magical operation conducted in 1946 by the rocket scientist John Whiteside 'Jack' Parsons and L. Ron Hubbard, the founder of the Church of Scientology (see Chapter 16). According to Nichols and Moon, the Babalon Working was 'an exhaustive operation which was designed to open an interdimensional door for the manifestation of the goddess Babalon (which means understanding), the Mother of the Universe'.

Like the Montauk Project, the Babalon Working was an attempt to open a doorway to another dimension, to master the secrets of space and time – in effect, to usurp the role of God. Nichols and Moon believe that the success of the Babalon Working resulted in 'infiltration from another dimension'; that the way was opened for an inimical alien force to enter our Universe. They cite the increase in UFO sightings around this time as evidence for their theory. Although this is inconsistent with the assertion made in their first book – that the UFOs entered our Solar System as a result of the deactivation of the device inside the DiPietro Molenaar Pyramid on Mars – it retains the central idea of some form of defence having been undermined. (Perhaps the deactivation of the pyramid device is what allowed the Babalon Working to succeed!)

Nichols and Moon include a chapter on 'Magick and Psychotronics' in *Montauk Revisited*, which throws some light on this confusing matter. They quote the magician Aleister Crowley's famous definition of Magick: 'the Science and Art of causing Change to occur in conformity with Will' (Crowley added a 'k' to the word to distinguish it from the bogus magic of conjurers). They see a similarity between Magick and science, in that both systems recognise that cause precedes effect, without reference to supernatural elements: 'the entire order and uniformity of nature underlie both systems'. While both systems are concerned with quantity (the empirical measurements upon which scientists base their theories), Magick extends beyond this into the idea of quality or the mysterious and unquantifiable nature of the human mind and its capacity to transcend mundane experience.

Although the idea of Magickal operations is used to cast light on the principles behind bizarre and outrageous stories like the Montauk Project, with its associated rumours of technological cooperation between humanity and alien beings, it also serves to explain how such elaborate scenarios arise. In *The Montauk Project*, Nichols and Moon include a 'guide to the reader' which describes the book as 'an exercise in consciousness'. A distinction is made between 'hard facts', which are backed up by physical or documentary evidence; 'soft facts', which are not untrue, but which are not amenable to verification; and 'grey facts', which are merely plausible. Of course, what one person calls plausible, another calls outrageous nonsense; and the story of the Montauk Project contains much that is plausible! This dichotomy between the concrete and the ambiguous is to be found throughout the vast and complex field of the paranormal.

In fact, Nichols and Moon say that their books may be read as science fiction, if it makes the reader more comfortable – an ambiguity, reminiscent of the *Illuminatus!* novels of Robert Shea and Robert Anton Wilson, which occupies a strange, nebulous realm between what is real and what is fiction. Whether or not it is true (and let us be quite honest, it probably isn't), the Montauk Project provides a focus for public apprehension regarding the perceived moral and ethical dangers that arise from scientific progress. Nichols' and Moon's books are underground bestsellers, and Nichols is a frequent guest at gatherings of conspiracy theorists with an interest in 'psychotronic' warfare and surveillance.

It should also be said that such conspiracy theories often contain elements drawn from popular culture, particularly science fiction. Anyone who has seen the film *Total Recall* (based on the short story 'We Can Remember it For You Wholesale' by the great American science fiction writer Philip K. Dick, who was himself prone to self-inflicted paranoid torment) will have noticed its parallels with the alleged alien machinery beneath the DiPietro Molenaar Pyramid on Mars. Nichols and Moon cleverly draw attention to the similarity, and claim that the film is 'fancifully based' on some of the events that occurred during the Montauk Project!

Conspiracy theories, whether they be political (the Kennedy assassination, the planned UN invasion of America, etc.), scientific (the secret testing of biological weapons on cattle), or merely bizarre (the Montauk Project), have undergone a constant metamorphosis over the decades. In fact, many elements within conspiracy theories have become interchangeable. For instance, mutilated cattle are seen by some people as the victims of human scientists engaged in biological weapons research, and by others as the victims of hungry aliens. The mysterious, so-called 'black helicopters' that are claimed to harass certain people are seen by some as an advance UN invasion force, and by others as UFOs in disguise. Some see the alleged alien presence on Earth as real, and others see it as an elaborate hoax to make us tolerate the formation of a repressive, one-world government.

Whether or not we follow Nichols' and Moon's suggestion, and regard the Montauk Project as science fiction, it is clear that it comprises virtually all the preceding theories. We have the secret manipulation of human beings, and the manipulation of history itself; we have alien involvement, irresponsible research into parapsychology and the unknown powers of the human mind; and we have the use of Magick in association with physics to delve into the mysteries of matter, space and time.

21

GLIMPSES OF OTHER TIMES

THE MYSTERY OF TIME SLIPS

If it is true, as theoretical physicists tell us, that there is no distinction between past, present and future, and that the past and future exist just as solidly as the present in the great fabric of spacetime, might it not be possible for us to catch glimpses of other times? Orthodox science would argue otherwise, claiming that our perception of time is fixed by the structure of our minds, that we are only capable of perceiving each moment as it passes, that the 'arrow of time' faces in one direction only, and that once a moment passes, it is gone from our perception for ever, existing thenceforth only as memory. The full capabilities of the human mind are far from understood, however, and mystics have long spoken of a non-material substance known as the Akashic Records, which contains an 'imprint' of every thought and event that has ever occurred. Is it possible that some people, wittingly or unwittingly, occasionally access these records, and as a result find themselves perceiving other times?

INTO THE PAST

On the morning of 23 October 1963, Mrs Coleen Buterbaugh was walking across the campus of Nebraska Wesleyan College towards the office of the Dean, Dr Samuel Dahl, whose secretary she was. At 8.50 precisely, she entered the C.C. White Building, where Dr Dahl had his office, and which was also used as a music hall.

Coleen walked briskly along the corridors, listening to the disjointed strains of music from practising students, then entered the office of Dr Tom McCourt, a visiting lecturer from Scotland. The office was part of a suite of two rooms, and Coleen was about to wish Dr McCourt a good morning before continuing through to her boss's office. As soon as she opened the door, however, she immediately became aware that something was not right. There was a curious, musty smell on the air; both rooms were empty and the windows were open.

She looked through to the second office, where another woman was standing with her back to her. The woman was looking into one of the drawers in a wall cabinet. Presently, Coleen realised that she could no longer hear the strains of music from the students: an eerie silence had descended, and she had the distinct impression that she was no longer part of that normal morning, as if she had somehow left it behind.

Coleen looked more closely at the woman in the second office, who seemed to be engaged in filing. She was tall, slim and dark-haired, and was wearing a long-sleeved white blouse and an ankle-length brown skirt, totally out of keeping with the fashion of the time.

In a conversation with a reporter named Rose Sipe of the Lincoln *Evening Journal*, Coleen later said: 'I still felt that I was not alone. I felt the presence of a man sitting at the desk to my left, but as I turned around there was no one there. I gazed out the large window behind the desk and the scenery seemed to be that of many years ago. There were no streets. The new Willard sorority house that now stands across the lawn was not there; nothing outside was modern.'

Coleen suddenly grew very frightened and left the room and the building. She could do no work for the rest of that day, and when she finally summoned the courage to tell Dean Dahl what had happened, much to her surprise, he was sympathetic, and took her to see another academic, Dr Glenn Callan, who was chairman of the division of social sciences, and asked her to repeat her story.

Dr Callan had been at the faculty since 1900, and after asking her several questions, he was able to conclude that Coleen had somehow 'seen' the office as it had been during the 1920s. The woman she had encountered was, Dr Callan suggested,

Miss Clara Mills, who had been head of the musical theory department. Was Miss Mills still alive? wondered Coleen. Dr Callan shook his head. Unfortunately, he said, Clara Mills was found dead in her office in 1930.

PARIS, 500 YEARS AGO

Occasionally, writers on mysterious and paranormal subjects experience profound and puzzling mysteries themselves. This happened to the highly respected naturalist and scientist Ivan T. Sanderson, who wrote extensively on mysterious animals, while he and his wife were living in the small village of Pont Beudet in Haiti, where they were conducting a biological survey along with Sanderson's assistant, Frederick Allsop.

One evening, the three of them decided to drive to Lake Azuey. On a remote dirt road, their car hit a patch of mud and sank up to its axles. There was nothing for it but to get out and walk. At one point, they managed to flag down a passing car; but the driver, an American doctor, apologetically told them that he was on his way to an urgent case, but added that he would pick them up on his way back. Sanderson and his companions thanked the doctor, and continued on in the moonlight. It was then that the most extraordinary thing happened.

According to Sanderson: 'Suddenly, on looking up from the dusty ground I perceived absolutely clearly in the now brilliant moonlight, *and casting shadows appropriate to their positions*, three-storied houses of various shapes and sizes lining both sides of the road.' The houses hung out over the road, which was muddy and covered with cobblestones. Sanderson guessed that the architecture was of 'the Elizabethan period of England', and yet, for some unknown reason, he felt absolutely certain that he was looking at a street in Paris. The houses had:

> . . . dormer windows, gabled timbered porticoes, and small windows with tiny leaded panes. Here and there, there were dull reddish lights burning behind them, as if from candles. There were iron-frame lanterns hanging from timbers jutting from some houses and they were all swaying together as if in a

wind, but there was not the faintest movement of air about us. I could go on and on describing this scene as it was so vivid: in fact, I could *draw* it. But that is not the main point.

The 'main point' was that Sanderson's wife saw all this, too! She had halted on the dirt track, and was gazing in wonder at the scene into which they had apparently stumbled. She was able to describe every detail, exactly as Sanderson was seeing it. When he asked her what she thought had happened to them, she replied with a question of her own: 'How did we get to *Paris* five hundred years ago?'

Presently, Sanderson and his wife began to experience weakness and dizziness, and they both called out to Allsop, who was some way ahead along the dirt track. He hurried back to them and, concerned for their well being, told them to sit down by the side of the track and offered them cigarettes. 'By the time the flame from his lighter had cleared from my eyes, so had fifteenth century Paris, and there was nothing before me but the endless and damned thorn bushes and cactus and bare earth.'

Fred Allsop had seen nothing unusual while walking along the track, and was deeply concerned at the apparent nonsense his friends were speaking. He suggested they just sit there and wait for the American doctor to return and pick them up, which after a while he did.

They arrived back at the Sandersons' residence early the next morning, and were surprised to find that their housekeeper had a hot meal waiting for them. She would not say how she had known they would be back at dawn; but later, one of the young men in the village said something very curious to Sanderson: 'You saw things, didn't you? You don't believe it, but you could always see things if you wanted to.'

Sanderson and his wife certainly *had* seen things – but how? Was it possible that a temporal 'window' had suddenly and inexplicably opened up between twentieth-century Haiti and fifteenth-century Paris? And why were they so sure that it was indeed Paris that they were seeing? As we shall see shortly, there may be an explanation – although it is not one upon which orthodox science would look very fondly.

A TELESCOPE INTO THE PAST

In his book *The Mammoth Encyclopaedia of Unsolved Mysteries*, Colin Wilson cites a case which was described to him by the person concerned, Mrs Jane O'Neill of Cambridge. In 1973, she was the first person on the scene of a serious traffic accident involving a bus, and helped many of the injured passengers to safety.

Some time later, on a visit to Fotheringay church, Mrs O'Neill was impressed by a painting of the Crucifixion behind the altar; but when, back at the hotel, she mentioned the painting to the friend with whom she was travelling, her friend said that there was no such painting in the church.

They revisited the church the following year, and Mrs O'Neill was surprised to discover that the interior was quite different from the way she remembered it, and of the painting behind the altar there was no sign. She subsequently contacted a local antiquarian and described what she had seen on her first visit to the church. He responded that she had seen the interior of the church as it had been in 1553.

Colin Wilson notes that the theory of 'time slips' was first developed in the mid-nineteenth century by two American professors, Joseph Rodes Buchanan and William Denton. Buchanan also coined the term 'psychometry', meaning the faculty which some people seem to possess by which they can somehow 'read' the history of an object simply by holding it in their hands. Denton tested his students with various geological specimens, which he wrapped in thick brown paper, so that they would have no idea what they were holding.

> A piece of lava brought 'visions' of an exploding volcano; a fragment of meteor conjured up visions of outer space; a piece of dinosaur tooth brought visions of primeval forests. Denton was convinced that all human beings possess this faculty, which he described as 'a telescope into the past'.

Another curious time slip occurred when a Mrs Turrell-Clarke was cycling along a road in Surrey. Without warning, she realised that she was no longer cycling, but *walking* along the road. As if this were not bizarre enough, she now was wearing nun's robes, and saw a man dressed in peasant clothes from the Middle Ages.

A month later, Mrs Turrell-Clarke was sitting in the village church in Wisley-cum-Pyrford, when she became aware that her surroundings had somehow reverted to their original state: the floor was covered with earth and the altar was made of stone. In addition, it was now filled with monks dressed in brown habits, who were singing the same plainsong that was being sung by the congregation in the modern church. Wilson writes: 'So it seems clear that what happened was that her viewpoint changed, and she found herself *looking through someone else's eyes* – the eyes of a lady walking along the road and the eyes of a woman standing at the back of the church.'

If this is true, then it may rather eerily account for the absolute certainty of Ivan Sanderson and his wife that they were walking along a street in fifteenth-century Paris: they knew this because they were looking at the street through someone else's eyes.

Many people who experience time slips speak of the feeling of 'crossing a threshold', a feeling that is accompanied by a profound silence. We are reminded of the sensation experienced by Coleen Buterbaugh as she entered the office on the Nebraska Wesleyan campus: the musical strains of the practising students, she said, faded away to be replaced by a heavy silence.

What is the nature of the 'threshold' crossed? It seems less likely that people who experience time slips physically enter a different time, than that their minds are temporarily displaced somehow, and they see other times through the eyes of people living in those times.

OUT OF THIN AIR

This would seem at least to be a working hypothesis; however, as so often with the subtle and complex field of the paranormal, there are cases on record which would seem to imply that something altogether more extreme and bizarre is happening. The realms of the unexplained never reveal easy answers.

For instance, on 6 January 1914, a naked man suddenly appeared in the High Street at Chatham, Kent, throwing the passers-by into a panic and scattering them like frightened rabbits. No one had seen him undress, and subsequent searches failed to locate his clothes. The man was apprehended by police and taken to the nearest police station, where he could tell them nothing about himself. Such was his demeanour that he was eventually declared insane.

More than half a century earlier, a man had been found wandering in a German village near the town of Frankfurt-an-der-Oder. He had no idea how he had come to be in the village, but said that he was from an unknown country called Sakria.

The latter case may well be no more than an item of folklore (how, for instance, was the man able to converse in German if he came from a country no one had – or has – ever heard of?). Altogether more intriguing is the discovery of a man in an immaculate suit, who was found dead below the west Botley flyover near Oxford. Apparently he had either fallen or been pushed from the flyover. Upon examination, it was found that the manufacturer's labels had been removed from his clothes, and he had no identifying documents on his person.

The only clues which were found were five handkerchiefs, each of which bore the monogram 'M', and a thin strip of foil containing 15 tablets. Later analysis showed the tablets to be a brand new drug called Vivalan – a drug so new that few doctors in England even knew of its existence. As the researcher Paul Begg notes, the man (who has never been identified) apparently stepped out of thin air, 'presumably at some point above the A420 road – and plummeted to his death'.

In 1920, a naked man was found dead, apparently from exposure, in a ploughed field near Petersfield in Hampshire. The incident was reported in the London *Daily News*:

> That a man could wander near to the main road between Petersfield and Winchester in a nude condition until he died in a field from exposure, aggravated by minor injuries such as cuts and abrasions, is astonishing, but that his identity and everything connected with his death should remain a mystery to-day is almost unbelievable.
>
> The man's nails were manicured, the palms of his hands showed that he was not engaged in manual labour, and his features and general appearance were

those of someone of a superior class. But although his photograph has been circulated north, east, south and west through the United Kingdom, the police are still without a clue, and there is no record of any missing person bearing the slightest resemblance to this man, presumably of education and good standing.

In 1975, a young man fell to his death from the 17-storey Kestrel House tower block in Islington, London. There were no clues to his identity beyond a pair of bus tickets and an envelope addressed to the National Savings headquarters in Glasgow. The police tried to follow up these scant clues, without success.

Inspector Robert Gibson of King's Cross police told the *Sunday Express*: 'Somebody somewhere must have loved him or at least known him.' Gibson could not believe that such a person could have gone missing without someone noticing and reporting it to the police. And yet no one did, and to this day the young man's identity remains a mystery.

Just as in the many cases of vanishing people discussed elsewhere in this book, it seems that people have, on numerous occasions, simply *appeared* out of thin air. Where have these people come from? When found, they are mostly either dead or in such a state of derangement that no meaningful information can be gleaned from them.

There are several possibilities. For one thing, they may have met with foul play of some kind – tragic, but mundane; or the survivors may have undergone some powerful (but earthly) trauma which they are too distressed to describe. There are many non-supernatural explanations – for some cases, that is. However, there remain other instances where the likeliest explanation seems to be that whatever happened to them was most certainly *not* natural, or at least not part of the 'nature' which we claim to understand.

If windows onto other times might open before the minds of some people on occasion, perhaps other types of portal can also open: physical portals through which the unwitting happen to step, and are snatched from their own time and unceremoniously deposited in ours.

Charles Fort comments on this matter in his book *Lo!*:

I suspect that many persons have been put away, as insane, simply because they were gifted with uncommon insights, or had been through uncommon experiences . . . If there have ever been instances of teleportations of human beings from somewhere else to this Earth, an examination of inmates of infirmaries and workhouses and asylums might lead to some marvellous astronomical disclosures . . . Early in the year 1928 a man did appear in a town in New Jersey, and did tell that he had come from the planet Mars. Wherever he came from, everybody knows where he went, after telling that.

22

COMET OR SPACECRAFT?

THE TUNGUSKA EXPLOSION OF 1908

M odern life is replete with dangers of every description, from AIDS to identity theft, from violent street crime to global warming and international terrorism. Statistically, some dangers are extremely unlikely, and yet the spotlight of media attention has illuminated them so starkly that they appear to loom over us like hungry predators waiting to pounce and destroy us at any moment. And there are other dangers which, while appearing to be even less likely, could spell the end of humanity if they ever came to pass. One such scenario is the collision of a large comet or asteroid with Earth. Many scientists believe that such an impact caused the extinction of the dinosaurs 65 million years ago, clearing the way for the rise of mammalian life and, ultimately, us. Another such impact could wipe *us* out, clearing the way for something else. With all the other things to worry about, most people prefer not to think about the possibility of a 'hammer of the gods' descending upon us from the depths of space; but should we be worrying more, and perhaps planning some form of preventative action?

At 7.17 on the morning of 30 June 1908, the Earth received a desultory flick on the nose from the Universe. That flick on the nose took the form of a column of blue light, as bright as the sun, which split the sky over central Siberia approximately 1,000 kilometres north of the town of Irkutsk and Lake Baikal. Tungus natives and Russian settlers in the region reported seeing the titanic flash and hearing a rumbling sound like artillery, followed by a blast wave that knocked them off their feet and shattered windows hundreds of kilometres away.

The first report of the explosion appeared in the Irkutsk newspaper on 2 July 1908, two days later. In part, it reads:

> The peasants saw a body shining very brightly (too bright for the naked eye) with a bluish-white light . . . The body was in the form of 'a pipe', i.e. cylindrical. The sky was cloudless, except that low down on the horizon, in the direction in which this glowing body was observed, a small dark cloud was noticed. It was hot and dry and when the shining body approached the ground (which was covered with forest at this point) it seemed to be pulverized, and in its place a loud crash, not like thunder, but as if from the fall of large stones or from gunfire was heard. All the buildings shook and at the same time a forked tongue of flames broke through the cloud.
>
> All the inhabitants of the village ran out into the street in panic. The old women wept, everyone thought that the end of the world was approaching.

An eyewitness in the village of Vanovara, S.B. Semenov, also provided a vivid and detailed description of what he experienced:

> I was sitting in the porch of the house at the trading station of Vanovara at breakfast time . . . when suddenly in the north . . . the sky was split in two and high above the forest the whole northern part of the sky appeared to be covered with fire. At that moment I felt great heat as if my shirt had caught fire; this heat came from the north side. I wanted to pull off my shirt and throw it away, but at that moment there was a bang in the sky, and a mighty crash was heard. I was thrown to the ground about three *sajenes* [about 7 metres] away from the porch and for a moment I lost consciousness . . . The crash was followed by noises like stones falling from the sky, or guns firing. The earth trembled, and when I

lay on the ground I covered my head because I was afraid that stones might hit it.

A farmer in the Kezhma region, about 200 kilometres south of the impact site, related his own experience of the blast:

> At that time I was ploughing my land at Narodima [6 kilometres west of Kazhma]. When I sat down to have my breakfast beside my plough, I heard sudden bangs, as if from gun-fire. My horse fell on its knees. From the north side above the forest a flame shot up. I thought the enemy was firing, since at that time there was talk of war. Then I saw that the fir forest had been bent over by the wind and I thought of a hurricane. I seized hold of my plough with both hands, so that it would not be carried off. The wind was so strong that it carried off some of the soil from the surface of the ground, and then the hurricane drove a wall of water up the Angara. I saw it all quite clearly, because my land was on a hillside.

The colossal explosion registered on seismographs at stations across the Eurasian continent, measuring 5.0 on the Richter scale. Through comparison of these data with seismograms of the Novaya Zemlya and Lop-Nor nuclear weapons tests, the physicist Ari Ben-Menahem determined that the Tunguska impactor had 'the effects of an Extraterrestrial Nuclear Missile of yield 12.5±2.5 megatons', about 30 times more powerful than the atomic bomb that destroyed Hiroshima, and about one-fifth as powerful as the largest-ever hydrogen bomb explosion.

According to recordings at meteorological stations in various countries, the air compression wave from the blast circled the planet twice. The temperature at the centre of the fireball was estimated to be in excess of 30 million °F [degrees Fahrenheit], and the blast uprooted and blew down approximately 80 million trees over an area of 2,150 square kilometres in the great taiga forest. Following the impact, vast forest fires broke out, and the effect of the impact on the atmosphere was such that for several weeks afterwards the night skies across Russia and Europe were bathed in an eerie light, bright enough to read a newspaper by at midnight.

Perhaps surprisingly (at least by our modern standards of rapid travel and instant news), there was no official recorded expedition to the region until 13 years later, in 1921. This is most likely due to the remoteness of the region and the lack of serious casualties (not counting, of course, the vast numbers of reindeer and other wildlife that were incinerated by the initial blast and the resulting forest fires). The first expedition was headed by Leonid Kulik, a mineralogist employed by the Soviet Academy of Sciences, who travelled throughout the region, interviewing eyewitnesses to the event.

From their testimony, Kulik concluded that a giant meteorite had crashed into Tunguska, and he managed to persuade the Soviet government to fund further expeditions to the region to search for the impact site itself.

In 1927, Kulik finally reached the site of the explosion in the Stony Tunguska River region. No amount of dry academic speculation could have prepared him for the sight which greeted him as he and his party climbed a ridge and gazed across the blasted landscape. Countless trees, some of them a metre in diameter, had been uprooted and cast to the ground like so many matchsticks. To their surprise, there was no evidence of an impact crater at the centre of the 50-kilometre-diameter zone of destruction, and even more intriguingly, the trees at 'ground zero' itself were still standing, although badly scorched, with their branches and bark stripped off.

During the following decades, there were several more expeditions to the region, none of which found any evidence to support Kulik's theory that the destruction had been caused by an impacting meteorite: there was no crater and no meteoritic remnants. Although the early expeditions had discovered holes in the landscape, these were later identified as natural depressions.

What, then, caused the blast? If not a meteorite, what could have laid waste such a vast area of land? What kind of object could cast 80 million trees to the ground, stripping them of their branches and bark, and yet leave no crater as a sign of its coming?

In 1930, the British astronomer F.J.W. Whipple tentatively suggested that the 'Tunguska event' might have been caused by a small comet. Comets are composed primarily of ice and dust, and such a body would have completely vaporised by contact with the Earth's atmosphere. Thus, although the energy released by such an event would have been colossal, no 'object' as such would have made direct contact with the planet's surface, and so there would have been no impact crater. This would also explain the eerily luminous night skies, which would have been caused by comet dust in the atmosphere refracting sunlight falling upon the Earth's dayside.

In a 1983 paper, the astronomer Zden k Sekanina criticised the comet hypothesis, claiming that a comet following the apparently shallow trajectory of the Tunguska impactor should have disintegrated very high in the atmosphere, whereas the object that caused the devastation in Siberia appears to have remained intact much closer to the ground. And yet, just over 20 years earlier, in 1962, microscopic pellets of magnetite and silicate globules were discovered in soil samples from the region. According to the Soviet astronomer V.G. Fesenkov, a double spherule consisting of a magnetite pellet inside a silicate shell could be the result of 'rapid condensation of incandescent gas upon cooling'.

Debate has continued for decades as to the true nature of the Tunguska impactor, with most orthodox scientists favouring either the asteroid or comet hypothesis, in spite of the problems each presents. However, some scientists (and many more researchers on the paranormal) have suggested other, more fantastic causes.

The first to do so was the Soviet science fiction writer and engineer Alexander Kazantsev, who speculated in a 1946 story that the Tunguska event might have been caused by a nuclear explosion on board a Martian spacecraft which was seeking fresh water from Lake Baikal.

Several decades later, this intriguing idea began to receive serious consideration by Soviet scientists, one of whom, Alexei Zolotov, suggested that the disaster might have been caused by the malfunctioning engines of a nuclear-powered alien spacecraft. Zolotov admitted, however, that there were serious problems with this hypothesis, not least of which was the likelihood that such an advanced vessel would doubtless have safety systems which would prevent such a catastrophe.

(We might also point out that, since all the other planets of the Solar System are uninhabited and incapable of supporting advanced life, such a craft would have to be interstellar, and it is extremely unlikely that its engines would be nuclear-powered: such a notion belongs more in the realm of 1940s science fiction than in the real world.) Zolotov added that the area of destruction was 'an amazing demonstration of pinpoint accuracy and humanitarianism'.

In his 1980 book *Stones From the Stars*, T.R. Le Maire speculates further on this notion, suggesting that 'the Tunguska blast's timing seems too fortuitous for an accident'. Even a slight change of course of the incoming object would have obliterated a major city. Le Maire suggests that 'the flaming object was being expertly navigated', using Lake Baikal as a reference point:

> The body approached from the south, but when about 140 miles from the explosion point, while over Kezhma, it abruptly changed course to the east. Two hundred and fifty miles later, while above Preobrazhenka, it reversed its heading toward the west. It exploded above the taiga at 60°55' N, 101°57' E.

Fascinating as the thought is of an alien spaceship crew acting altruistically to guide their doomed vessel away from the Earth's population centres, to die alone and unmourned above the wastes of an alien planet, there is no evidence that the Tunguska impactor changed course in the final moments of its fiery flight. Certainly, none of the eyewitnesses mentioned such a curious event.

Other unusual and fascinating theories have been put forward to explain what happened at Tunguska. In a 1965 article in the journal *Nature*, C. Cowan, C.R. Alturi and W.F. Libby suggested that a piece of anti-matter from deep space might have undergone an annihilation reaction with the atoms in the Earth's upper atmosphere. Anti-matter carries an opposite electrical charge to matter (for instance, an electron carries a negative charge, and its anti-matter equivalent, the positron, carries a positive charge), and when they come into direct contact, the two annihilate each other and are converted to energy with 100 per cent efficiency. (In contrast, a nuclear explosion converts matter to energy with approximately 4 per cent efficiency.)

Intriguing as this theory is, however, there is little, if any, evidence to support it, and quite a lot to contradict it: for instance, if such pieces of anti-matter were floating around in interplanetary and interstellar space, they would constantly be reacting with the particles of gas and dust with which space is filled. We would then be able to detect the energetic gamma rays produced in these annihilation reactions, and to date we have been unable to do so. In addition, Cowan *et al.* suggested that such an event would have resulted in a significant increase in levels of Carbon-14 in the surrounding vegetation; and yet no such increases have been observed in the Tunguska region.

If an anti-matter micro-meteorite can be discounted as the cause of the Tunguska event, might some other equally exotic denizen of the mysterious gulfs of space be to blame? A black hole, for instance?

In 1973 University of Texas physicists Albert Jackson and Michael Ryan suggested that a very small black hole, with a mass of 10^{22} to 10^{23} grams, would have the necessary energy to have caused the devastation in the Tunguska region. Such an object, although incredibly tiny, would have sufficient mass to slice through the Earth like a hot knife through butter, ploughing through the body of the entire planet before exiting through the Atlantic Ocean on the opposite side, causing a similarly catastrophic release of energy.

Once again, fascinating as this theory sounds, it is most unlikely to be true. According to W.H. Beasley and B.A. Tinsley, writing in *Nature* in 1974, the recently invented microbarographs in Britain, which recorded the airwaves caused by the blast in 1908, did not record any subsequent airwaves issuing from the Atlantic Ocean, which would have been an inevitable consequence of the exit of the black hole. Nor does the black hole hypothesis explain the magnetite and silicate globules mentioned above, which were found in the region of the blast; and it certainly doesn't account for the brightness in the night sky which occurred for several weeks after the event. As Beasley and Tinsley conclude: 'All the evidence favours the idea that the impact which caused the Tunguska catastrophe involved a body with characteristics like a cometary nucleus rather than a black hole.'

Their conclusion is further borne out by the work of a team of Italian researchers, which recently visited the region in an attempt to solve the Tunguska mystery once and for all. Combining an analysis of seismic records from several Siberian monitoring stations with data on the direction of flattened trees, the team managed to calculate the orbit of the Tunguska impactor. According to the team's leader, Dr Luigi Foschini, they 'performed a detailed analysis of all the available scientific literature, including unpublished eyewitness accounts that have never been translated from the Russian'.

Their conclusions point to an object approaching Tunguska from the south-east at approximately 11 kilometres per second; from this they were able to plot a series of possible orbits for the object. According to the BBC science journalist David Whitehouse, 'Of the 886 valid orbits that they calculated, over 80% of them were asteroid orbits with only a minority being orbits that are associated with comets. But if it was an asteroid,' he asks, 'why did it break up completely?'

According to Foschini, the object may have been like the asteroid Mathilde, which was photographed by a space probe in 1997. 'Mathilde is a rubble pile with a density very close to that of water. This would mean it could explode and fragment in the atmosphere with only the shock wave reaching the ground.'

While this scenario is certainly plausible, a curious appendix to the nuclear explosion hypothesis was introduced in 1989 when some astronomers suggested that some of the deuterium in a comet entering the Earth's atmosphere may have undergone a nuclear fusion reaction, which resulted, in effect, in a natural nuclear explosion.

In spite of the excellent work done by many scientists over the decades since the Tunguska catastrophe, it seems that its exact nature will be the subject of intense speculation for many years to come – or until such a catastrophe occurs again, affording us a tragic opportunity to observe it in greater detail ...

23

STARSHIP

IS FASTER-THAN-LIGHT TRAVEL POSSIBLE?

THE GREAT DREAM

It is one of the greatest dreams of humanity: to plunge into the Galaxy's depths and explore the vast whirlpool of stars, in which our Sun and its family of worlds are the merest flecks – hardly more than dust motes drifting through the unending stellar night. To climb aboard a starship and begin the ultimate adventure is, of course, a commonplace in science fiction, so much so that many science fiction writers prefer to leave it aside and explore other aspects of humanity's possible future.

The fascination of fiction writers with spaceflight actually pre-dates what we now think of as 'science fiction': as long ago as 1865, Jules Verne described a hypothetical lunar flight in *From the Earth to the Moon*, and, of course, H.G. Wells wrote of a lunar expedition in *The First Men in the Moon* in 1901. The early imaginings of spacecraft were naïve in the extreme: most were bullet-shaped or otherwise streamlined – a superfluous design feature in the empty void of space.

The idea of spaceflight gained respectability following the end of the Second World War. The devastation visited upon London by the German V-2 ballistic missiles proved that such technology was practicable, and many writers wondered what could be achieved if such technology were to be developed and used for peaceful purposes rather than aggression.

The idea of space exploration as something more than fantasy, something feasible that might one day come to pass, took root in the public imagination thanks to the work of science fiction novelists and short story writers like Robert Heinlein (1907–88), who also wrote the screenplay for the classic George Pal film *Destination Moon* (1950), and Arthur C. Clarke, whose novel and screenplay for *2001: A Space Odyssey* is as visually stunning and intellectually intriguing today as in 1968, when the film was released. These and many other stories of near-future exploration of the Solar System did much to expand the public consciousness, making people aware of the immensity of space and the boundless potential of humanity's future away from our planetary cradle.

Science fiction, however, is a vast genre (most of its concerns and tropes are well beyond the scope of this chapter), and its musings on the future of humanity in space extend far beyond the Solar System. From its early phase in the first decades of the twentieth century, science fiction was not content merely to explore the nearby planets of our own Solar System: after all, there was an entire galaxy out there, and countless other galaxies beyond. Many science fiction writers decided that the really interesting stuff was happening far out in the unexplored interstellar gulfs.

Chief among these early pioneers was E.E. 'Doc' Smith (1890–1965), whose 1928 novel *The Skylark of Space* began one of the most famous and celebrated of science fiction epics, the Skylark Series. These stories, it must be said, have dated appallingly, and today are read mainly for either nostalgia's sake, or with an amused and ironic 'post-modern' eye. Heroines who faint every time an alien pops up, and starships that reach faster-than-light speeds by simple acceleration (more on this later) are unsophisticated to say the least, and have not stood the test of time at all well. 'Doc' Smith (the 'Doc' refers to his PhD in Chemistry) went on to produce another hugely famous and influential series of novels, the Lensman Series, in which the whole of human civilisation has been manipulated by two warring super-civilisations, the benign Arisians and the malignant Eddorians.

These and other stories come under the sub-genre heading of 'space operas'. The term was coined with reference to 'soap operas', and was intended as a dismissive putdown; however, the sub-genre subsequently gained acceptance and respect, and today it is healthier than ever (if the phenomenal success of *Star Wars*, *Star Trek* and countless other films, TV shows and books is anything to go by). The early space opera tales told exuberantly of interstellar exploration and conquest, with massive space battles in which entire star systems were annihilated, and took it for granted that humanity would spread relentlessly among the stars, establishing great interstellar empires. In the early days, these were based very much on the Roman Empire; perhaps the most famous of these 'future histories' is Isaac Asimov's hugely influential Foundation Trilogy, which he later expanded into a longer series of novels.

In more recent years, the space opera has been given a new lease of life by the work of such hugely successful writers as Iain M. Banks (author of the 'Culture' series of novels) and Peter F. Hamilton (author of the Night's Dawn Trilogy). As David Pringle writes in *The Ultimate Encyclopaedia of Science Fiction*: 'The universe beyond the solar system now seems a far stranger and more hostile place than it did in the days when "Doc" Smith . . . imagined it as a giant playground for fanciful boys' games, but the prospect of exploring and colonizing the wilderness of stars still produces a powerful effect on the science-fictional imagination, and will doubtless continue to do so.'

All of these wondrous and wonderful tales are predicated on a single piece of technology, which has gone by many names over the decades. It is variously known as the hyperdrive, the stardrive, the jumpdrive, the warp drive and so on. It has a single function, without which space opera and many other sub-genres of science fiction would not be possible: to propel a spacecraft faster than the speed of light.

THE PROBLEMS OF STARFLIGHT

Faster-than-light (FTL) interstellar travel is a great dream, to be sure; but how close are we to making it a reality? Many scientists maintain that FTL travel will always remain a dream. We will never be able to call down to our engineers and demand 'warp factor nine'; our engineers will never be able to fret over their superluminal engines, warning us that the ship will explode if we push her any further.

It is indeed impossible for a spacecraft to travel faster than the speed of light (186,000 miles per second) by means of known propulsion systems. This is due to relativistic mass increase, which means that the faster a body travels, the greater its mass becomes. Although Albert Einstein's Theory of Special Relativity established him as the central figure in twentieth-century physics, the relations which he used to explain the behaviour of space, time and matter within a moving reference frame were originally derived by a Dutch physicist named Hendrik Lorentz. For this reason, they are called the Lorentz transformations. In his 1988 book *Faster Than Light: Superluminal Loopholes in Physics*, Dr Nick Herbert summarises the Lorentz transformations:

> The Lorentz transformations predict four major changes that objects in a moving frame seem to undergo as viewed from a fixed frame: (1) Space shrinks in the direction of motion (Lorentz contraction); (2) time slows down (time dilation); (3) clocks desynchronise (relativity of simultaneity or 'sync shift'); and (4) mass increases in the moving frame.

Herbert goes on to explain that at low speeds (compared to that of light), the Lorentz transformations are insignificant. However, as the speed of light is approached, the transformations increase by an amount proportional to the 'Einstein factor'. 'The Einstein factor is equal to 1 for low velocities, is equal to 2 for 90 per cent of light speed, increases to 7 at 99 per cent of light speed, and becomes infinite at light speed itself.'

Thus, the mass of a spacecraft accelerating towards the speed of light would increase, presenting increased resistance to further acceleration. Not only would the energy of the propulsion system be usurped by the increasing mass, but at the speed of light itself, the spacecraft's mass would become infinite; therefore, the amount of energy required to reach the speed of light would also be infinite. In short, the ship would have to carry an infinite amount of fuel, capable of delivering an infinite amount of thrust!

To give some idea of the size of this problem, Herbert describes the behaviour of electrons inside a television set, which are accelerated to 30 per cent of light speed. For these electrons, which 'paint' the picture you see on your screen, the Einstein factor is about 1.05, and they are 5 per cent heavier than electrons outside the television.

Electrons in a particle accelerator, on the other hand, move a great deal faster – so fast that their Einstein factor is 50,000. A bullet travelling as fast as an electron in a particle accelerator, says Herbert, 'would weigh as much as a dump truck'.

POSSIBLE SOLUTIONS

It's clear that we will never attain FTL travel through conventional methods of propulsion. We will have to look elsewhere for clues as to how we might one day bypass the apparently impassable light barrier and journey to the stars.

The mysterious – one might say magical – realm of quantum theory might just hold such clues, since among other things it describes a curious phenomenon known as 'quantum tunnelling', in which particles can move from one location to another without traversing the space between them. Another notion is the 'quantum connection' which suggests, in Herbert's words, that 'once two quantum systems have briefly interacted, they remain in some sense forever connected by an instantaneous link – a link whose effects are undiminished by interposed shielding or distance'.

This continued instantaneous interaction seems to occur as a result of the mathematical probability waves which represent the quantum systems, and which do not exist in normal three-dimensional space, but rather in what is called 'configuration space', which contains three dimensions for each particle. According to Herbert: 'The quantum wave for a two-particle system, for instance, moves about in a six-dimensional space.'

Herbert likens this faster-than-light quantum connection to the workings of voodoo, in that an action on one particle instantly affects the other particle, because they leave parts of themselves with each other during their initial interaction, and are still in contact with each other through those parts.

Many science fiction writers have used the notion of the 'tachyon' in their FTL stardrives. Herbert summarises the properties of these hypothetical particles, and the differences between them and tardyons (slower-than-light particles) and zero-mass luxons (particles that travel at the speed of light, only three of which are currently known: protons, neutrinos and the hypothetical gravitons, which carry the force of gravity):

- A tachyon possesses an imaginary rest mass; that is, the square of its mass is a negative number.

- As it loses energy, a tachyon speeds up . . . Once a tachyon has lost all its energy, it must travel at an infinite velocity; that is, it occupies every point along its trajectory at the same time. Olexa-Myron Bilaniuk and E.C. George Sudarshan, authors of an early treatise on tachyons, call a particle that dwells in this strange state of omnipresence – zero energy/infinite velocity – a 'transcendent' tachyon.

- To slow a tachyon down requires the addition of energy. To slow a tachyon down to light speed requires an infinite quantity of energy. Thus, for a tachyon, the speed of light is a lower limit to its velocity. Once a tachyon, always a tachyon – such particles can never go slower than light.

If a way could somehow be found to 'translate' the matter of a starship into tachyons, it could then travel faster than light – indeed, it wouldn't have much choice! Of course, this may well be impossible since, presumably, its rest mass would also have to be converted to an imaginary number, which would result in the starship becoming an 'unphysical entity'.

THE ALCUBIERRE WARP DRIVE

Another possible method of FTL travel was suggested by the Mexican theoretical physicist Miguel Alcubierre in 1994. It starts from Einstein's realisation that matter warps the spacetime around it. Alcubierre was interested in whether the fictional 'warp drive' used in *Star Trek* could ever be realised. This led him to search for a valid mathematical description of the gravitational field that would allow a warp in the fabric of spacetime to serve as a means of superluminal propulsion.

Alcubierre concluded that a warp drive could be feasible if matter could be arranged so as to expand the spacetime behind a starship (thus pushing the departure point many light years back) and contracting the spacetime in front (bringing the destination closer), while leaving the starship itself in a locally flat region of spacetime, inside a 'warp bubble' lying between the two distortions. The ship would then surf along inside its bubble at an arbitrarily high velocity, pushed forward by the expansion of space at its rear and the contraction of space in front. It could travel faster than light (*much* faster, in fact) without breaking any of the laws of physics because, with respect to the spacetime inside its warp bubble, it would actually be at rest. Also, being locally stationary, the ship and its crew would not be troubled by any devastating accelerations and decelerations, and from relativistic effects such as time dilation.

Could the 'Alcubierre Warp Drive' ever be built? As Alcubierre himself points out, it would require the manipulation of matter with a negative energy density. Such matter, which is known as 'exotic matter', is the same kind of strange stuff needed to maintain stable wormholes, another theoretical means of circumventing the light barrier, and one also beloved of science fiction writers.

Although Alcubierre's theory sounds outlandish, it is based on the laws of physics as we currently understand them (even though the actual construction of his starship is still way beyond our current technology). Inspired by the Mexican physicist's proposal, other physicists have begun to take the idea of FTL travel more seriously, and this has resulted in a new branch of theoretical physics – what might be called 'warp drive theory'.

It may take centuries for us to figure out how to build Miguel Alcubierre's warp drive-propelled starship (if we *ever* figure it out); but, as the great Chinese philosopher Lao Tzu said, a thousand-mile journey begins with a single step . . .

24

VISITORS FROM THE UNIVERSE?

THE ORIGIN OF UFOS

No book on scientific mysteries would be complete without a chapter on the UFO phenomenon. However, there are so many books already available which catalogue sighting after sighting (which, I hasten to add, is no bad thing), it may be better to devote this chapter to a regrettably but necessarily brief examination of the theories that have been put forward to account for the sightings, both of UFOs and their occupants, that have occurred over the decades.

THE EXTRATERRESTRIAL HYPOTHESIS

In spite of the many theories that have been presented over the years to account for UFO sightings, by far the most popular is still Extraterrestrial Hypothesis (ETH), which holds that unexplained encounters with UFOs and their 'occupants' represent the activities of one or more scientific expeditions from another planet. Although some researchers maintain that this explanation is among the least viable, it has become firmly entrenched in the public psyche, and is still supported by many of the most-respected UFO researchers.

To understand the reasons for this, we must go back to the early days of ufology, when humanity was just beginning to take its first tentative steps beyond the confines of Earth. Science had enabled us to control our environment, to create a life of relative ease and prosperity (at least in what was then called the First World), especially in the United States. In the years following the Second World War, this prosperity increased with the continued advances in technology and industrial production, allowing America not only to improve the lives of its own citizens, but to finance the regeneration of Europe through the Marshall Plan. The atomic bombing of Hiroshima and Nagasaki had demonstrated humanity's potential to master the secrets of matter itself, not to mention its ability to destroy itself.

Science and technology, then, constituted the apogee of human endeavour, in terms of both creation and destruction. While Robert Oppenheimer referred to himself as 'death, the destroyer of worlds', the power of the atom was also seen as the liberator of the human race, a source of abundant energy that would allow us to fulfil completely our potential as masters of this planet. These miraculous advances led us to consider science as central to our relationship with reality.

When faced with a set of mysterious events, therefore, we naturally view them within a scientific context. When Kenneth Arnold encountered nine crescent-shaped objects in the skies above Mount Rainier in Washington State on 24 June 1947, the initial assumption was that he had seen a group of high-performance aircraft being secretly tested by either the United States or some foreign government, possibly the USSR. The Air Materiel Command came forward with a more mundane explanation: the 'objects' were either reflections of the Sun on low clouds, small meteorites breaking up in the atmosphere, or large flat hailstones. However, several scientists dismissed these explanations as nonsense. This was the genesis of public suspicion of an official cover-up.

It was also the beginning of a spate of sightings of unconventional objects displaying flight capabilities far beyond anything achievable at that time (or even today). The daylight sightings tended to fall into two categories: flat, disc-shaped objects and larger, cigar-shaped objects. Both types seemed to be metallic, structured craft under intelligent control, but without visible means of propulsion – no propellers or jet engines, and no wings or other stabilising structures. Assuming that these craft were indeed of unknown origin, and neither hoaxes nor misidentifications of mundane objects such as birds, meteorites, other aircraft or planets (all of which, along with many other events, undeniably account for more than 90 per cent of UFO sightings), the most acceptable conclusion was that the strange objects were not built by humans. They had arrived from elsewhere.

Science had already demonstrated its boundless potential in a number of different ways; it was clear that humanity was just starting its journey towards greater and greater knowledge. It was surely not too outrageous to suggest that, in the far reaches of space, there might be many cultures which had progressed much further on their own journey, and that their thirst for knowledge had brought them to our planet.

When people across the world began to report encounters with the supposed pilots of these craft, the ETH became even more firmly cemented in the public imagination, despite the exasperated protestations of both sceptics and serious ufologists, who were deeply suspicious of such alien-encounter claims. Even national newspapers, while maintaining a sceptical stance, were more than willing to publish such accounts.

At this stage, the ETH was still a new theory, one which seemed to many to be perfectly logical: the history of science is littered with shattered paradigms – assertions that had seemed incontrovertible until a new idea came along to explain observed reality in a more successful and complete way. To the proponents of the ETH, this was (and still is) the case with UFOs. The Ptolemaic view of the Universe, in which all the heavenly bodies move around the Earth, had given way to the Copernican view, in which the Sun is at the centre of the Universe. This, in turn, was modified over the centuries, until humanity came to realise that it occupies an insignificant planet orbiting an insignificant star in an average galaxy in a medium-sized galactic cluster in a Universe that may or may not be the only one.

The somewhat conceited notion that the Earth harbours the only intelligent species (and perhaps the only life) in the Universe is still a matter of intense debate; to proponents of the ETH, it is a further example of a paradigm that was made to be broken – and *has been* broken by the presence of UFOs and their occupants. But in view of what we have discovered about evolution and genetics, how feasible is the ETH as a theory to explain the presence of UFOs and apparently alien creatures?

It has long been maintained by orthodox science that UFOs cannot be spacecraft from another star system. For one thing, the distances involved in interstellar travel are simply too vast. Moreover, nothing can travel faster than the speed of light (approximately 186,000 miles per second); it is the absolute limit to how fast a spacecraft (or anything else) can move. Thus, interstellar travel becomes an incredibly costly and time-consuming proposition, with journey times – even to the nearer stars – running into millennia. According to the sceptics, it follows that UFOs cannot be alien spacecraft, since no civilisation would be either willing or able to make such an enormous investment in time and resources. In fact, this is the main rationale for SETI (the Search for Extraterrestrial Intelligence), whose supporters argue that it would make much more sense for advanced civilisations to make contact through radio or other transmissions, rather than sending large, energy-hungry spacecraft.

However, science (like most human endeavours) seems to be divided into those who say something can't be done, and those who go right ahead and do it anyway. As Arthur C. Clarke said, scientific breakthroughs tend to occur in four phases: one, it is impossible and will never be done; two, it is possible but far too expensive ever to be done; three, I always said it was entirely possible; and four, I thought of it first!

Apparently, it is the same with the concept of faster-than-light interstellar travel. As we saw in the previous chapter, theoretical physicists are now postulating the use of gravity amplifiers to contract the space in front of a vessel, while elongating the space behind it, thus allowing it to cover vast distances in a tiny fraction of the time it would take a vessel travelling through normal space. Of course, this is just a theory, and such spacecraft haven't even reached the drawing board yet; but today's theories sometimes turn out to be tomorrow's realities, and it seems that the time is fast approaching when the sceptics will no longer be able to cite the 'light barrier' as confirmation that UFOs cannot come from other star systems.

While the ETH is theoretically possible, we find ourselves immediately encountering another stumbling block: the outrageously high frequency of sightings. In his book *Revelations: Alien Contact and Human Deception*, the respected UFO researcher Jacques Vallée gives an initial figure of 5,000 reported close encounters with UFOs, stretching back over the last 40 years or so. Taking account of the fact that only about one in ten close encounters are reported, together with the phenomenon's global nature, the geographical distribution of reports in terms of population density, plus the nocturnal patterns that seem so prevalent, Vallée extrapolates the number of UFO landings that have actually occurred in the last 40 years – in other words, the actual number of reports that would be received if everyone in the world could remain in a constant observational state, without going to bed at night, and so on.

The total comes to approximately 14 million.

Vallée goes on to ask what possible motive alien beings could have for landing on a planet 14 million times in four decades. He reminds us that unlike Venus, for instance, the Earth's surface is visible from space, making mapping the planet the simplest of procedures. The visitors could also monitor our radio and television broadcasts in order to collect a colossal amount of cultural data, without interacting with us at all. Should they require physical specimens of soil, flora, fauna, etc., it would be possible to land unobtrusively and collect them. While this would probably entail making several landings, it would surely not entail making *14 million*.

Another very serious problem with the ETH is that of the aliens' physical appearance. Although some rather bizarre shapes have been reported over the decades (including giant brains and eyeless beings with tentacles and only one leg), the vast majority have borne a striking resemblance to humans, with a recognisable head, two arms, two legs and a torso. Indeed, the so-called 'Grey' alien has now become accepted, both in the media and by a significant number of ufologists, as the standard model of what genuine extraterrestrials look like.

Proponents of the ETH cite this as evidence that the humanoid configuration is the optimum for intelligent species throughout the Universe. For example, it makes a lot of sense for the two eyes (providing stereoscopic vision) to be located on the head, as close to the brain as possible. It also makes sense for the head to be on top of the body, providing the best possible view of the being's environment. Two hands with opposable thumbs are best suited to the fashioning and use of tools, while two legs ensure swift and effective locomotion (more than two would take up valuable processing capacity in the brain).

This sounds reasonable enough, until one remembers that the humanoid shape is a very finely tuned adaptation to conditions on this planet, and this planet alone. Also, it is the result of tens of thousands of random mutations in our genetic material, occurring over millions of years; mutations that, at any time in our history, could have occurred in a subtly different way, taking us in a different evolutionary direction. Aliens originating on a distant planet would be the result of evolutionary processes within their native biosphere, which would almost certainly have different characteristics to those of Earth. For instance, it is highly unlikely that the aliens' planet would have exactly the same mass as Earth, with an atmosphere of exactly the same density and with identical gases in identical proportion. It is possible that beings with four limbs, a backbone, a large skull and manipulative hands could evolve on another planet, but they would still be strikingly different from us. Random genetic mutation, combined with the diverse conditions that must prevail on other worlds, makes it extremely unlikely that intelligent extraterrestrials would look anything like us.

Of course, while the ETH cannot account for unexplained close encounters in general, that does not mean that it cannot account for *some* of them. Take the case of the Florida scoutmaster D.S. Desvergers, for example, who, along with his troop, witnessed a UFO landing in some nearby woods. Leaving the boys behind, Desvergers went into the woods to investigate. He became aware of a disc-shaped object hovering above him. The disc sprayed a fiery substance at him, which burned his arms. There was a 'turret' on top of the object, inside which Desvergers saw a creature so horrifying that he could not bring himself to describe it. If the scoutmaster was telling the truth about what happened to him, the occupant of the UFO could conceivably qualify as a genuine non-human extraterrestrial.

ULTRATERRESTRIAL EVOLUTIONARY CONTROL SYSTEMS

As Jacques Vallée is fond of saying, the Extraterrestrial Hypothesis is not strange enough to account for the evidence of non-human encounters. There have been far too many landings for a planetary survey; this colossal number is as ludicrous as the reported behaviour of the UFO occupants, who frequently communicate with humans in absurd ways. (Vallée feels that 'absurdity' is not quite the right word to describe the entities' behaviour, and prefers the term 'metalogic'.)

For instance, in France in 1954, a man encountered a UFO. The occupant asked him what time it was. The man looked at his wristwatch and replied: 'It's 2.30.' The UFO occupant said: 'You lie – it is four o'clock.' However, the time *was* 2.30. The occupant then asked: 'Am I in Italy or Germany?'

These two apparently absurd questions, one about time, the other about space, don't seem quite so absurd on closer examination. Vallée wonders whether these exchanges might represent the UFO occupant's attempt to instruct the human witness on relativity, to tell us that time and space are not what we think they are. There is an interesting parallel here with the famous Barney and Betty Hill UFO abduction case from New Hampshire, USA, in 1961. When Betty Hill told the 'leader' of the aliens who had captured them and conducted physical examinations of them that her husband, Barney, had false teeth because he was getting old, the alien didn't understand, and Betty found herself trying to explain the concept of time, with which the beings were apparently unfamiliar. However, when the Hills were about to leave the alien craft, the leader said to Betty: 'Just a minute ...', implying that they *were* familiar with the concept of time. Although this has been interpreted as an internal inconsistency in the Hills' story, implying a shared dream as the likeliest explanation of their experience, we can speculate that there may be a deeper meaning, one that does not necessarily preclude the dream hypothesis.

It is possible that a very subtle, symbolic communication took place aboard the UFO, in which the superficiality of the human understanding of time was demonstrated. A close examination of these exchanges reveals our assumption that the 'aliens' do not understand the concept of time to be erroneous. They were playing a game with Betty Hill, leading her to believe something that was not true, and only giving the game away (doubtless intentionally) at the very end of the encounter. In other words, the encounter tells us that we do not know what we think we know, either about the 'aliens' or about the Universe itself.

It is this symbolic, almost playful, element to non-human contact that has led writers like Jacques Vallée and Whitley Strieber to suggest that we may be dealing with the activities of some kind of 'control system', which is guiding our religious, philosophical and cultural development in ways that are both immensely subtle and virtually impossible to comprehend in terms of our current scientific paradigms. Whitley Strieber has said that his so-called 'Visitor Experience' could be what the forces of evolution look like when applied to conscious minds.

Strieber has also said that the visitors appear to be interacting with humans on a direct, one-to-one basis, completely bypassing the routes one might expect an alien expedition to follow in making contact with another culture. The leaders of our world, whether political or religious, do not seem to be of the slightest interest to the visitors, who much prefer to appear in the lives of ordinary human beings.

There is no doubt that encounters with these beings are often extremely traumatic. Even if a percipient is spared the nightmarish medical procedures reported by so many, the simple fact that an obviously non-human intelligent being is standing before them is enough to send a shudder of abject terror through the soul. Strieber cites stress tests that were carried out on rats in the 1970s. The rats were made to suffer electric shocks for long periods, with the result that they grew stronger and more intelligent: their suffering actually improved them (recalling the dictum that whatever doesn't kill you makes you stronger). Strieber goes on to say that, as he remembered this and realised that the function of the visitors is somehow to force us to evolve, a 'tired, young voice' said: 'Thank you.'

If we are being forced to evolve by a higher intelligence whose true nature remains hidden from us, the question remains: what is the ultimate goal of this evolution? Many people who have claimed contact with 'extraterrestrials' have been profoundly altered by their experiences. Whatever the appearance of the beings themselves, the results of contact are similar.

They include the acquisition of paranormal abilities such as precognition, telepathy and psychokinesis (the movement of objects through the use of the mind alone), and also a sudden, consuming awareness of a wider spiritual realm, echoing the Platonist view that what we call reality is no more than a flimsy veil, hiding a far more profound yet accessible eternity. They become intimately concerned with the fate of the Earth, something which may have occurred to them only fleetingly before their experiences.

DENIZENS OF OTHER DIMENSIONS?

Some UFOs are reported to have the ability to appear or vanish instantaneously, and also to plunge into and emerge from solid ground. There are a large number of reports in which UFOs seem to change shape, as if they are not behaving according to the accepted laws of physics. One of the most impressive examples of this occurred on 29 June 1954. The BOAC Stratocruiser *Centaurus* was flying from New York to Newfoundland, when the crew saw a large, metallic object off the port side of the aircraft. Surrounding the large UFO were six smaller objects, apparently escorting it. The captain of the Stratocruiser began to sketch the large object, which, astonishingly, seemed to be changing shape, from a delta-wing, to something that looked like a telephone handset, to a pear shape.

A Sabre jet fighter was scrambled from Goose Bay, Newfoundland, to investigate. The Sabre pilot radioed that he was in range, and that he had a radar lock on the *Centaurus* and the UFOs. As he approached, however, the six smaller objects lined up in single file and merged with the large one, which then diminished in size and suddenly disappeared.

Although this activity could indicate an extraterrestrial understanding of physics vastly surpassing our own, when viewed in conjunction with the metalogical behaviour of some UFO occupants, another, stronger implication comes to mind – one that points to a far more complex set of events and circumstances than straightforward alien visitation. The evidence seems to suggest an *extradimensional* rather than extraterrestrial origin for UFOs and their occupants.

The ability of UFOs (and non-human beings, for that matter) to wink in and out of existence has some strange implications, which are increasingly being supported by our own advances in the fields of high-energy and theoretical physics. There is a distinct possibility that our strange visitors might be denizens of a 'superspace' in which our own Universe is embedded, like a single bubble in a multi-dimensional foam.

The American researcher John A. Keel, who has been called 'the last of the great ufologists', presented his theory of 'ultraterrestrials' in his book *UFOs: Operation Trojan Horse*. Keel suggests that it is possible that the UFOs and their occupants are actually natives of the Earth – not this Earth, that is, but an Earth existing at the same coordinates in a parallel spacetime continuum. In support of his theory, Keel cites not only the capability of UFOs to appear and disappear at will, but also the colours they display while doing so. He reminds us that ultraviolet light immediately precedes the visible colour spectrum, the first visible frequencies being purple or violet light. If UFOs have their origin in a dimension of a different electromagnetic frequency to ours, as they entered our dimension they would appear as indistinct, purplish blobs of light. Once within our spacetime continuum, they would continue to 'gear down from the higher frequencies', their colour changing from ultraviolet to violet, and then to cyan, or blue-green. Once in our continuum, they would radiate energy on all frequencies, resulting in a harsh, white glare. Keel writes:

> In the majority of all landing reports, the objects were said to have turned orange (red and yellow) or red before descending. When they settle to the ground they 'solidify', and the light dims or goes out altogether. On takeoff, they begin to glow red again. Sometimes they reportedly turn a brilliant red and vanish.

This hypothesis would seem to be supported by the observational data that have accumulated over the years, regarding changes not only in the colour but the shape of UFOs. However, the apparently absurd, or metalogical, behaviour of the UFO occupants towards humans implies that, although there is a physical phenomenon at work here, it also extends into the realm of the human psyche, and is perhaps influenced by its interactions with humanity in a two-way exchange. The phenomenon clearly takes account of the prevailing worldview in the society with which it is interacting, and alters its surface characteristics to appear as angels and demons, or fairies, or alien explorers.

Although quantum theory is frequently drafted in to account for all manner of paranormal phenomena, it is worth considering the idea that ufonauts, if they really are denizens of a realm beyond our spacetime continuum, may be a function of the 'Hidden Variable' theory developed by the physicist Dr David Bohm. According to this theory, quantum events are determined by a sub-quantum system acting outside spacetime. And although space and time as we know them do not exist in this realm, it is nevertheless the origin of all phenomena occurring in this and all other universes. If the Hidden Variable is actually consciousness (as some physicists suggest), we can see at least a tentative link between reported interactions with a non-human intelligence, and a scientifically sound mechanism by which such an intelligence could exist. Indeed, the realm of the psyche might be the best (perhaps the only) theatre of interaction between human beings and a non-human intelligence located outside this spacetime continuum.

This 'ultraterrestrial' intelligence could have its origin in one of the infinite parallel realities predicted by the Many Worlds interpretation of quantum physics, according to which all of the potential outcomes of every event occurring in this Universe result in a branching-off into an alternate universe. All of these universes are equally 'real', and exist in their own parts of a higher 'super-spacetime'. It is conceivable that a sufficiently advanced civilisation might have mastered the techniques of 'jumping' between these realities in order to follow its own unknown agenda.

If consciousness is as important as some physicists suggest in their interpretation of the Hidden Variable theory, Carl Jung's theory that UFOs represent our fundamental desire for wholeness and unity might be given a new lease of life. As Vallée suggests in *Revelations*, 'the human collective unconscious could be projecting ahead of itself the imagery which is necessary for our own long-term survival beyond the unprecedented crises of the twentieth century'.

VISITORS FROM THE FUTURE?

An interesting alternative to the extraterrestrial and ultraterrestrial hypotheses, and one that is fairly acceptable to those who believe that UFOs are solid 'nuts-and-bolts' vehicles, is the Time Travel Hypothesis (TTH). This suggests that the UFO occupants are actually our own distant descendants, who have travelled back through time to this era, and probably many other eras in the past.

The TTH also accounts for the physical similarities between ourselves and the UFO occupants which, as mentioned earlier, should militate against their extraterrestrial origin. It might also explain the occasionally reported cooperation between the diminutive, spindly-limbed 'Greys' and another group of entirely human-looking ufonauts, known as 'Nordics'. Perhaps humans from the relatively near future are engaged in joint ventures with those from the distant future.

This is all very well, but is time travel actually possible? Time travel, like interstellar travel, is a commonplace in science fiction, and is still considered by many to be a physical impossibility. But, like rapid interstellar travel, it has received more serious attention from theoretical physicists in recent years. There are a number of possibilities that might, one day, lead to the construction of a practical time machine.

For instance, spacetime is composed of four dimensions: three of space (left/right, forward/backward, up/down) and one of time. The three spatial axes extend at right angles to each other, and the time axis *also* extends at right angles to the three spatial axes. This is extremely difficult (if not impossible) to visualise, but it can be described quite easily in mathematical terms. If a machine could be constructed that was capable of amplifying gravity to a sufficient extent, the time axis could be warped to the extent that it becomes a spatial axis. Time travel might then become possible.

Intense gravitational fields also form the principle behind a theoretical device with the unlikely sounding name of 'Tipler's Infinite Rotating Cylinder'. In 1974 the American physicist Frank Tipler suggested that a sufficiently massive cylinder, rotating at half the speed of light, could distort spacetime to the extent that 'closed timelike loops' are created, through which past time would become accessible. British physicist

John Gribbin has calculated that Tipler's cylinder would have to be 100 kilometres long, 10 kilometres in radius, have a mass equivalent to the Sun and the density of an atomic nucleus, to work.

There is, however, a problem with building such a device (quite apart from the obvious one): the rotating cylinder would be so dense that it would instantly undergo axial collapse, and would end up looking more like a giant pancake than a cylinder. It may be forever beyond our capabilities to build Tipler's Infinite Rotating Cylinder, although Gribbin suggests that it may one day be possible to utilise natural objects, such as rapidly rotating pulsars, for the same purpose.

Another reason why time travel may be impossible is that nature may well refuse to allow us to create temporal paradoxes. This is a well-known concept that, once again, has an extensive provenance in the world of science fiction. Time paradoxes violate a putative law known as the Causal Ordering Postulate (COP), which insists that causes must precede effects. The most well known of these paradoxes is that in which a psychopathic time traveller goes back into the past to kill his own father before he (the time traveller) was conceived. Many physicists maintain that time travel will always be impossible because a scenario such as this would violate the COP. Of course, one needn't choose such an extreme case to illustrate the time paradox. There is a story by Ray Bradbury in which time travellers go sightseeing in remote prehistory, and return to find their democratic world transformed into a totalitarian hell. One of their number discovers the reason: a crushed butterfly on the sole of his shoe. Even a seemingly insignificant alteration in the past might have far-reaching consequences, which the COP would not allow.

However, all is not lost for the time-travel enthusiast. It may just be that quantum theory has a way to avoid creating paradoxes. According to the Many Worlds interpretation, we do not live in a 'Universe' as such, but in a 'Multiverse' composed of an infinite number of 'ghost' realities existing alongside this one. The Many Worlds interpretation was arrived at through experiments with single photons (particles of light), but can be applied to events in the macroscopic world. In John Gribbin's words, we live amidst 'a myriad array of ghost realities corresponding to all the myriad ways every quantum system in the entire universe could "choose" to jump'. So, in the macroscopic world, every time we make a choice (to cross a road at a particular point, for instance), the universe divides into an infinite number of slightly different 'branches' in which we made a different choice.

If the Many Worlds interpretation is an accurate description of reality, then the problem of time paradoxes would not arise. Any irresponsible time traveller intent on altering the past would simply create an alternate universe, quite separate from his own, which would thus be preserved. He could still kill his own father prior to his conception, but (in this sense at least) it wouldn't matter: it would just mean that in the new universe, he was never born. In his native universe, however, he *was* born.

Science fiction writers have come up with many other elegant solutions to the problem of temporal paradoxes. For instance, in Michael Moorcock's scientific romance *The Dancers at the End of Time*, the characters are prevented from causing paradoxes by time itself: whenever they are on the point of doing so, the time in which they happen to be spits them out like pieces of bad food, hurling them unpredictably into a different time.

It can thus be seen that the initial groundwork, at least, for a workable theory of time travel has been laid. Whether it ultimately proves to be correct, only time will tell (if the reader will pardon the expression).

The TTH might also account for the apparently absurd behaviour of the UFO occupants, in particular their harvesting of human genetic material, including sperm and ova. We can imagine a future in which the decline of human fertility (which is already giving cause for concern) will reach the point where the very survival of the species is under threat. Faced with the prospect of extinction, it might well make sense for our descendants to travel back into history to collect the materials essential to their survival.

However, we are still on rather shaky ground here, since there would be no need for the time travellers to abduct people: all they would have to do is break into a few fertility clinics, where they would find all the sperm and ova they would need to bolster their genetic stock. It is perhaps more likely that they would use their time travelling abilities to perform historical research, in the search for knowledge that has characterised human activity throughout its history.

We must also take into account the possibility that there may be insurmountable difficulties in sending material objects through time. But sending *information* through time might be perfectly manageable, if use could be made of fundamental particles such as the hypothetical tachyons, which travel faster than light, and hence from the future into the past. This might well account for the ghostly behaviour of UFOs and their occupants. Perhaps they are not physical objects and beings at all – at least, not in this time – but data projections from holographic transmitters in the future, capable of recording information from the present and then transmitting it back to their own time.

If we speculate even more wildly on this theme, we find ourselves approaching John Keel's theories again. He suggests that UFOs are composed of an energy that, once it has entered our reality, can be manipulated to 'temporarily simulate terrestrial matter'. If our remote descendants could transmit such an energy back through time – an energy that could be manipulated to simulate matter and thus interact more fully with people and objects in the present – this would account for the physical traces of their visits which have been reported on occasion.

POWER POINTS AND PSYCHIC VORTICES

One of the favourite explanations put forward by sceptics to account for UFO sightings and encounters with non-humans is that they are 'all in the mind'. To some extent, this is a perfectly acceptable statement, since the vast majority of UFO sightings are indeed misinterpretations of mundane objects viewed in unusual conditions. For instance, an airship catching the rays of the setting sun, and seen from an odd angle, might very well resemble a strange, glowing sphere that slowly changes its shape to that of an ellipse as the wind catches it and makes its profile more visible. The human mind has a tendency to interpret visual images in a way that makes sense. If we see something unusual in the sky, something for which there is no immediate and obvious explanation, our minds try to interpret it as something we can understand, so that even stars and planets can be seen as structured, artificial objects.

In an elaborate extension of this explanation, researchers such as Paul Devereux and Michael Persinger have suggested that the movements of rocks below the Earth's surface sometimes produce electromagnetic phenomena in the form of so-called 'Earth Lights', glowing balls of light which have been reported as UFOs. Human beings in the vicinity of such phenomena might be subject to realistic hallucinations caused by the electromagnetic stimulation of the hippocampus region of the brain, which results in an altered state of consciousness. The visions and physical sensations associated with these altered states have been duplicated under laboratory conditions, through the electrical stimulation of the relevant structures in the brain.

Earth Lights often appear near power lines, transmitter towers, isolated buildings, roads and railway lines. It seems that certain cultures have always been aware of Earth Lights, and have even incorporated them into their belief systems. According to the prolific writer on folklore and the paranormal, Rosemary Ellen Guiley, the Native American Snohomish people of Washington State regarded them as doorways to other worlds, while the Yakima (also of Washington State) used them in their divination rituals. The Australian Aborigines consider Earth Lights to be the manifestation of the dead, or of evil spirits.

The area around the town of Sedona, Arizona, has gained prominence in recent years as one of the most active of these places. Situated about 40 miles south of Flagstaff, the area is sacred to the Yavapai people, who believe that Sedona's red rocks are home to various deities. Sedona is said to contain a 'power point': a place containing supernatural energies as yet undiscovered by science, but corresponding to the planetary consciousness of the Earth.

Paranormal phenomena such as UFOs, Earth Lights and apparitions of various kinds are frequently reported in the vicinity of power points. Some percipients also report acquiring psi (parapsychological) faculties, such as clairvoyance, or out-of-body experiences. Guiley sees a possible link between geomagnetic field activity and some psi activity in humans, thus suggesting that the phenomena are caused by the energies at the sites.

The implication here is that there is some sort of interaction occurring between the energy emanating from the power points, and the so-called 'bioenergy' postulated by the maverick scientist Wilhelm Reich (1897–1957). The concept of bioenergy has formed the basis of parapsychological research in Eastern Europe for many decades. Russian researchers claim that such phenomena are observable in the laboratory, and they even claim to be able to store bioenergy.

Eastern Europe is not the only place where research into these subjects is taken seriously. Yoichiro Sako is a senior researcher for the Sony Corporation of Japan. Although a graduate of the prestigious Tokyo University, Sako is not quite what one would call an 'orthodox' scientist. In 1991, he was appointed director of Sony's ESP (extrasensory perception) laboratory, a special research department directly approved by the Sony founders Akio Morita and Masaru Ibuka. Sako describes the field of ESP as 'a new technological revolution', and although he has yet to come up with any ideas as to how ESP could be incorporated into marketable products, the very fact that his department is being budgeted means that Sony believes the research has potential.

However, Sony has come under attack from some scientists for their funding of ESP research. Yohishiro Otsuki, a professor of physics at Waseda University, said he was furious at the corporation for financing an employee who believes in paranormal phenomena. Sako countered by declaring: 'We need to have a . . . holographic vision. Our ultimate goal is to discover the "mind or consciousness" that all humanity, and the whole of creation, must possess.'

Sako claims to have obtained evidence of psychic ability. One experiment involved placing a piece of platinum in one of two black film canisters, leaving the other empty. When asked which of the canisters contained the platinum, 'normal' people guessed correctly 50 per cent of the time; but a certain 'super-sensitive' person named 'T.I.' guessed correctly 70 per cent of the time.

Sako also echoes the idea of the 'sheep/goat' effect, first identified in the 1940s by the American parapsychologist Gertrude Schmeidler, which demonstrates that people who believe in psychic abilities tend to score more highly in tests than people who are sceptical. Sako maintains that these phenomena 'never happen among the deniers, no matter how long you wait'. For this reason, he believes that only people who are predisposed to a belief in paranormal phenomena are truly qualified to conduct useful research into them.

These concepts and experiments have been drawn together under the discipline known as 'psychotronics', which aims to study the ways in which matter, energy and consciousness interact with each other. According to proponents of the discipline, all paranormal phenomena arise from the vital force generated by all life in the Universe.

The first 'psychotronic generators' were developed by the Czech inventor Robert Pavlita, who claimed to have been inspired by certain ancient manuscripts, which he refused to name. The generators were composed of bits of machinery, humanoid figures, writing utensils and Easter Island monoliths. With them, Pavlita claimed to be able to store energy collected from any biological source, and used it to enhance plant growth, purify polluted water and kill insects.

FURTHER INTO PARAPSYCHOLOGY

If we are prepared to allow UFOs and their occupants to slip yet further into the realm of parapsychology (an idea from which many ufologists would recoil with horror), we can find an intriguing correlation with the phenomenon known as 'apports'; and this might shed a little more light on the mechanism by which they are manifested. The Hungarian psychoanalyst and paranormal researcher Nandor Fodor defined apports as 'the arrival of various objects through an apparent penetration of matter'. Although most apports are small objects such as coins, pebbles, rings and so on, and are (allegedly) usually manifested during spiritualist séances, larger and more complex objects, such as live animals, are occasionally reported.

Mediums say that apports are gifts from the realm of the spirits. While some researchers have suggested that the objects have their origin in other dimensions, from which they are taken by the medium through his or her psychic willpower, others suggest that the medium takes objects from various locations in this dimension and teleports them to the séance, disintegrating them at their initial location, and then reassembling them. (Of course, the principle of Occam's Razor would suggest that trickery is rather more likely!)

Teleportation of objects by human beings is said to be accomplished through the manipulation of universal energy (corresponding to the bioenergy of Wilhelm Reich). When spiritual adepts reach certain states of ecstasy, their bodies allegedly free themselves from the pull of gravity, allowing them to levitate and move at impossible speeds. Likewise, the apports produced by mediums occasionally float and move around, a phenomenon that has been linked to some poltergeist cases.

The phenomenon of apports could thus account for reports of UFOs and their occupants, in that they are described as appearing suddenly, performing wild aerobatic manoeuvres, and then disappearing as if into thin air. The objection of many ufologists that the objects sometimes leave physical traces is countered by substitution of the term 'paraphysical' for 'physical'. Air Marshall Sir Victor Goddard, who coined the term 'ufology' in 1946 while he was on the Allied Chiefs of Staff Advisory Committee in Washington, DC, observed that physical traces such as patches of burnt grass seem 'almost never to be there the next day when all that made them come to human consciousness of men like us has gone'.

Goddard believed this to imply that UFOs are apports, originating in:

> . . . a world pervading ours, co-incident with ours – its denizens created by astral mind-imagining. That which they create can manifest materially as apports in a transient state of hardware here. If so, the hardware *will* leave marks where it has landed on the ground, before it flies away again, or is again etherealised.

UFO and non-human encounters may thus be an external result of interactions between the human psyche and an as-yet incomprehensible life force permeating the Universe, and generated by all living entities within it. The ultimate purpose of these encounters remains unclear, but it is possible that one of their principal functions is to guide humanity towards a new phase in its evolution.

That such encounters should result in increased psi functioning in percipients should not come as a great surprise in this context. If UFOs really are apports – solid physical (or paraphysical) objects originating in the super-consciousness of the Universe itself – such improvements in the psychic abilities of percipients might constitute an inevitable, and perhaps essential, alteration in consciousness as a tool with which to apprehend the Universe and our place in it.

The power points mentioned earlier might even serve as amplifiers for these paranormal signals. If we really are constantly bathed in this bioenergy, generated by ourselves and every other living thing in the Universe, we may be ill-adapted to notice it at our present stage of evolution. We might be in a similar position to someone listening to an old gramophone whose sound horn is missing, struggling in vain to catch the tinny fragments of music rising from the record. The power points scattered all over the world might function in a way similar to that of an electronic amplifier, boosting the signal so that the frustrated music-lover listening to the gramophone is suddenly plunged into the ecstatic sweep of a crystal-clear symphony.

Perhaps this is what humanity has always been striving towards, without the majority even realising it. The pursuit of knowledge in all its forms – scientific, artistic, spiritual – is an attempt to forge, or perhaps rediscover, the link between the self and the infinite Universe that gave it life. Perhaps the ongoing encounters between humanity and non-human beings are reassurances that someone or something, somewhere, is responding to our efforts.

25

ZERO POINT ENERGY

A SCIENTIFIC HOLY GRAIL

As the old saying has it, there's no such thing as a free lunch. And yet we have never lost the dream of tapping a source of energy that is infinite and comes at little or no cost. The quaint ideas for perpetual motion machines with which our ancestors struggled (and which always ended up violating at least one law of thermodynamics) have undergone a renaissance in a new development in physics, which just might realise the dream of free energy.

It's called zero point energy (ZPE), and although we know that it definitely exists, scientists are divided on just how much of it there is, and how much of it can be tapped. According to modern physics, the idea that a vacuum is nothing more than empty space is quite wrong. It churns and roils with invisible activity, even when the temperature is absolute zero – the lowest temperature that can ever be attained. At absolute zero (–273.15°C), atoms and molecules have the minimum amount of energy allowed by quantum theory; at absolute zero, all molecular motion ceases.

No one knows how much zero point energy resides in the vacuum of spacetime, but some cosmologists have speculated that at the beginning of the Universe there was a huge amount – enough even to have triggered the Big Bang. Today, the amount is much, much smaller, although some scientists believe there is still more than enough to supply us with all the energy we could ever conceivably need . . . if only we could discover how to tap into it.

So what *is* zero point energy? The concept derives from a long-accepted idea in quantum theory (which describes the behaviour of fundamental particles like atoms, protons and electrons), which is known as Heisenberg's uncertainty principle.

Like many aspects of quantum physics, the uncertainty principle is bizarre and difficult to reconcile with the macroscopic everyday world; it states that the position and momentum (a combination of mass and velocity) of fundamental particles are linked in such a way that they cannot both be precisely measured at the same time. The principle is named after the great German physicist Werner Heisenberg (1901–76), who formulated it in 1927. As John Gribbin states in his *Companion to the Cosmos*: 'The key feature of quantum uncertainty is that it has nothing to do with the ability (or inability) of our instruments to make accurate measurements; it is an intrinsic property of the quantum world.'

If the position of a fundamental particle is known precisely, then its momentum is completely unknown, and vice versa. This is the origin of zero point energy: at absolute zero, particles *must* still be moving; they cannot be completely still, because, if they were, both their position and momentum would be known with absolute accuracy. Since this would violate Heisenberg's uncertainty principle, it cannot be so.

The uncertainty principle doesn't just apply to position and momentum: it applies to energy and time also. Even more astonishingly, it allows particles and bubbles of energy to (quite literally) appear out of nothing – as long as they disappear again within a very short period of time. Although occurring at infinitesimally small scales, the effects of ZPE can be measured in the laboratory.

THE CASIMIR EFFECT

Named after the Dutch physicist H.B.G. Casimir, who discovered it in 1948, the Casimir effect occurs when two metal plates, brought sufficiently close together, begin to attract each other very slightly. In an article in the December 1997 issue of *Scientific American*, Philip Yam explains what happens in the Casimir effect. 'The reason [for the effect] is that the narrow distance between the plates allows only small, high-frequency electromagnetic "modes" of the vacuum energy to squeeze in between. The plates block out most of the other, bigger modes. In a way, each plate acts as an airplane wing, which creates low pressure on one side and high pressure on the other. The difference in force knocks the plates toward each other.'

The actual force generated by the Casimir effect is, however, extremely tiny, and 'corresponds to the weight of a blood cell in the earth's gravitational field', according to the physicist Steven Lamoreaux, who devised an incredibly sensitive apparatus to measure the Casimir effect.

If we are ever going to tap ZPE successfully, we will have to look elsewhere than the Casimir effect; and this is precisely what some researchers are attempting to do.

Notable among them is Dr Hal Puthoff of the Institute for Advanced Studies in Austin, Texas (not to be confused with the world-renowned Institute for Advanced Studies in Princeton, New Jersey). Since the late 1980s, Puthoff and his colleagues have been testing various devices with the intention of finding a way to make ZPE a viable proposition for useful energy generation.

One of the most intriguing devices Puthoff tested was based on the phenomenon of 'sonoluminescence', whereby water is bombarded with sound waves until tiny air bubbles are created, which then collapse and give off flashes of light. This phenomenon occurs when the collapsing bubble creates a minute shock wave, which heats its interior suddenly, whereupon it reaches a flashpoint, releasing a minute amount of light.

In this device, the surfaces of the bubbles are said to act as the 'Casimir force plates', with the excluded vacuum energy converted to light. However, in spite of grandiose claims made for this device – that it generates more energy than is put into it and thus could become a provider of 'free energy' – Puthoff and his colleagues have been unable to do so under laboratory conditions.

Nevertheless, Puthoff is confident that one day ZPE will be harnessed in sufficient quantities to offer a solution to the problem of humanity's ever-increasing energy requirements. Moreover, he believes that the answer may lie in the structure of the atom. According to quantum theory, the electrons in an atom possess a minimum, ground state energy, which keeps them from spiralling into the nucleus and collapsing the atom, even though they lose energy through radiation. Puthoff suggests that it is actually ZPE which keeps the electrons in orbit around the atomic nucleus: in other words, they are constantly absorbing the ZPE, thus balancing their own energy loss through radiation.

HOW MUCH 'FREE ENERGY' IS THERE?

Potentially, at least, the mathematics of quantum mechanics suggests that in any given volume of empty space there is an infinite supply of 'vacuum energy frequencies', and hence an infinite supply of energy (proponents of ZPE are fond of pointing out that the volume of space in a coffee cup contains enough vacuum energy to boil all the oceans on Earth). Potentiality is not the same as actuality, however; and the fact remains that physicists dislike the notion of 'infinity', and try to get rid of it whenever it turns up in their calculations.

Orthodox physicists maintain that the actual amount of ZPE in any given volume of space is incredibly small, and thus of little use to our energy-hungry civilisation. In support of their assertion, they point to the nature of spacetime itself, which is flat (at least in the absence of massive objects). Because matter and energy are equivalent (as expressed in Einstein's oft-quoted equation $E=mc^2$), and matter exerts a gravitational force, physicists maintain that a vacuum packed with massive amounts of energy would likewise generate a massive gravity field, which would distort spacetime in a way that is simply not observed, and would have resulted in a very different Universe from the one we see around us.

Nevertheless, ZPE proponents like Hal Puthoff are confident that one day, we will find a way to utilise ZPE – and not just in the generation of usable energy, but also in other fields such as spacecraft propulsion. He suggests that inertia (the reluctance of an object to change the way it is moving) is a result of the 'drag effect' of moving through the zero point field. Since this field can, in principle, be manipulated, it may be possible to reduce a spacecraft's inertia, and hence to reduce its requirement for fuel and propulsive force.

If this is ever achieved, then perhaps one day the 'inertialess drive' created by the science fiction writer E.E. 'Doc' Smith might become a reality. If that were to be achieved, then accelerating a spacecraft to the speeds necessary for rapid and routine interstellar travel would be as simple and straightforward as driving to your local supermarket.

26

THE FINAL EXPERIMENT

COULD WE ACCIDENTALLY
DESTROY THE WORLD?

A MULTITUDE OF THREATS TO LIFE ON EARTH

The renowned British astronomer Martin Rees has suggested that humanity may have only a 50 per cent chance of surviving the next 100 years. The dangers facing our world are manifold. They lurk in the depths of space in the form of giant, planet-killing asteroids and comets; further out in the cosmos, exploding stars could shower the Earth with lethal radiation, effectively sterilising the planet so that not even bacteria would survive. The potential dangers posed by dying stars are serious enough, but those posed by *dead* stars are orders of magnitude greater: if a black hole were ever to drift into our Solar System, the Earth and every other planet, and even the Sun itself, would be consumed – ripped apart and swallowed by the ravenous singularity at the black hole's centre.

Although very real, these cosmic threats are unlikely in the extreme, and belong more in the realms of apocalyptic science fiction than in our day-to-day lives. However, there are other threats facing humanity which demand serious consideration. Global warming is the most obvious one, since climate change in its severest form could spell the end of our technological civilisation, and at the very least may well result in the death, suffering and displacement of millions of people – perhaps billions – in the centuries to come.

Others point to nano-technology as a potentially lethal threat to our species, citing a possible scenario in which self-replicating microscopic machines might escape from our control, running amok, consuming all life and covering the entire surface of the Earth with a suffocating 'grey goo'. Computers, too, constitute a potential threat to our civilisation. Although this sounds bizarre, even counter-intuitive (after all, computers and information technology have improved our lives immeasurably, and have resulted in a world undreamed of by our ancestors), the possibility remains that the superintelligent machines of the future may look upon their creators with something less than goodwill, and may decide that they can run the world better and more efficiently than human beings (and judging by the mess we have made of things so far, who is to say that they would be wrong?).

Bio-terrorism is another factor to consider. It is an unfortunate fact that several nations have possessed advanced biological weapons capability for many decades, and the United States and the United Kingdom (to mention just two) have spent many years perfecting the design and dispersal of lethal pathogens as part of their ongoing attempts to improve counter-measures against biological attacks.

In his frightening and fascinating book *Our Final Century*, Martin Rees writes that in the 1970s and '80s, the Soviet Union 'was engaged in the largest-ever mobilisation of scientific expertise to develop biological and chemical weapons'. Kanatjan Alibekov was a high-ranking scientist in the Soviet *Biopreparat* program during that period, following which he defected to the United States in 1992. In his book *Biohazard*, Alibekov (who changed his name to Ken Alibek) recounts how he was in charge of 30,000 workers at the *Biopreparat* laboratories, and how he was charged with the task of modifying various organisms to make them more virulent and more resistant to vaccines. In 1979, 66 people died mysteriously in the city of Sverdlovsk; and it was only in 1992 that Boris Yeltsin admitted that these deaths were the result of the accidental release of anthrax spores from one of the laboratories.

It is not such a huge leap of the imagination to suggest that some of this lethal material may well fall into the wrong hands at some point. Rees reminds us that in September 2001, envelopes containing anthrax spores were sent to two US senators and to several media organisations, resulting in five deaths. 'One can readily envisage the massive consequence for the national psyche of an outrage that killed thousands,' writes Rees, and adds that the 'actual impact of a future attack could be greater if an antibiotic-resistant variant of the bacterium were used, and, of course, if it were dispersed effectively'.

To be sure, human civilisation faces some dire threats, some more likely than others. However, there is one possibility that is potentially not only earth-shattering, but literally cosmos-shattering in its implications. Thankfully, it is also among the least likely of all the doomsday scenarios facing us; but the risk is still above zero ...

DOOMSDAY EXPERIMENTS

In *Our Final Century*, Rees reminds us of the mathematician and mystic Blaise Pascal's famous argument in favour of devout behaviour. Even those who consider it extremely unlikely that a vengeful God exists would still consider it prudent to behave as though He did, because it would be 'worth paying the (finite) price of foregoing illicit pleasures in this life as an "insurance premium" to guard against even the smallest probability of something infinitely horrible – eternal Hellfire – in the afterlife'.

Rees cites this as an extreme version of the 'precautionary principle', in which it is considered advisable to proceed cautiously in new areas of scientific research in spite of the promise of improvements to our lives which they make. A good example of this is the genetic modification of plants and animals: Rees maintains that 'the onus should be on the advocates of genetic modification to convince the rest of us that [our] fears are ungrounded – or, at the very least, that the risks are small enough to be outweighed by some specific and substantial benefits'.

This is indeed what happened during the Manhattan Project to develop an atomic bomb during the Second World War. The physicist Edward Teller, who worked on the project, wondered whether the detonation of the bomb would start a chain reaction which would ignite the Earth's atmosphere or oceans. In a Los Alamos report written before the Trinity test of the first atomic bomb in 1945, Teller examined a scenario involving a 'possible runaway reaction of atmospheric nitrogen', and wrote of his concerns that the 'safety factor' decreased rapidly with the initial temperature following the bomb's detonation. This concern was voiced again in the 1950s with the development of the hydrogen bomb, which generates even higher temperatures than the atomic bomb.

The physicists' calculations showed that the 'safety factor' was extremely large, and we now know that they were correct (since we are all still here!). But even though the danger of atomic and hydrogen bomb tests igniting the atmosphere was in fact incredibly small, it was still significant enough to give pause to the physicists concerned.

The same can be said of modern physics research, which attempts to create conditions more extreme than any seen since the very first moments of the Universe. It has been suggested that such experiments could, conceivably, unleash an uncontrollable chain reaction which resulted in the destruction not only of the Earth, but of the entire Universe.

Physicists try to understand how atoms and their component particles work, how they interact with each other and, ultimately, how they originated in the Big Bang, the primordial explosion that gave rise to the Universe. In order to probe the mysteries of matter and its constituent particles, physicists need to accelerate the atoms to speeds approaching that of light, which they do in vast particle accelerators, and then guide them into head-on collisions using powerful magnetic fields. The resulting collisions smash the atoms apart, revealing their internal structures.

According to Rees: 'An experiment that generates an unprecedented concentration of energy could – conceivably but highly implausibly – trigger three quite different disaster scenarios.'

The first scenario is the accidental creation of a miniature black hole in the laboratory. Black holes are formed in the Universe when a star reaches the end of its life and collapses in upon itself. If the star is of sufficient mass (several times that of our own Sun), it will continue to collapse, eventually becoming an infinitely small, infinitely dense 'singularity', from which nothing – not even light – can escape, such is the colossal intensity of its gravity. According to some scientists, there is a risk that such an object could inadvertently be created inside a laboratory on Earth, and would then proceed to consume everything around it, including, ultimately, the planet itself.

However, at present the energy required to create a black hole is millions of times greater than that produced in even the most powerful of our particle accelerators; and it seems to be extremely unlikely that humanity would ever gain access to energy levels sufficient for this scenario to become a reality. In addition, such microscopic black holes would almost certainly be harmless, in as much they would be *so* small that they would evaporate away almost instantaneously in a puff of so-called 'Hawking radiation', named after the great cosmologist Stephen Hawking, who discovered it.

The second scenario was brought to the British public's attention by a report in *The Times* on 18 July 1999. In part, the report stated:

> A nuclear accelerator designed to replicate the Big Bang is under investigation by international physicists because of fears that it might cause 'perturbations of the universe' that could destroy the Earth . . .
>
> Brookhaven National Laboratories (BNL), one of the American government's research bodies, has spent eight years building its Relativistic Heavy Ion Collider (RHIC) on Long Island in New York state . . .
>
> Last week . . . John Marburger, Brookhaven's director, set up a committee of physicists to investigate whether the project could go disastrously wrong.
>
> It followed warnings by other physicists that there was a tiny but real risk that the machine, the most powerful of its kind in the world, had the power to create 'strangelets' – a new type of matter made up of subatomic particles called 'strange quarks'.

Quarks are one of the two families of particles (leptons being the other) from which all known matter is composed; each proton and neutron in an atom contains three quarks. Strange quarks have been observed in the laboratory before, but they have always been attached to other particles. If a strangelet, composed of unattached strange quarks, were produced in the RHIC on Long Island, there is a small risk that it could convert anything it touched into more 'strange matter'.

Rees gives us the analogy of 'ice nine', the dangerous substance in Kurt Vonnegut's satirical science fiction novel *Cat's Cradle*. Ice nine is a new form of ice that is solid at room temperature. A global catastrophe is caused when a sample of ice nine is released from the laboratory and comes into contact with normal water. The result is the rapid solidification of all the water on Earth. 'Likewise,' writes Rees, 'a hypothetical strangelet disaster could transform the entire planet Earth into an inert hyperdense sphere about one hundred metres across.'

The third scenario is by far the most bizarre and catastrophic. According to some physicists, spacetime can exist in more than one state, or 'phase', just as water can exist in three phases: ice, liquid and vapour. Moreover, the vacuum of spacetime which encompasses our Universe may not be the 'true vacuum', but a fragile and unstable 'false vacuum'. Once again, Rees offers us a useful analogy, that of 'supercooled' water. 'Water can still cool below its normal freezing point if it is very pure and still; however, it takes only a small localised disturbance – for instance, a speck of dust falling into it – to trigger supercooled water's conversion into ice.'

It might be that the Universe is in a stable, 'supercooled' state, and that we are not living in the 'true vacuum' at all. It's also conceivable that high-energy collision experiments could 'rip a hole' in the fabric of spacetime, triggering a vacuum phase transition. Beginning in the laboratory, the boundary between the spacetime we know and this new, 'true vacuum', this new and strange form of reality, would propagate outwards at the speed of light, an expanding bubble of total oblivion.

In less than one-twentieth of a second, the Earth and everything on it would be engulfed by the new form of spacetime, and would be utterly annihilated. A little over one second later, the Moon would likewise disappear in the destructive nothingness of the new spacetime. Eight minutes later, the Sun would vanish, and a few hours after that, the rest of the Solar System.

But the destruction would not end there. Like Vonnegut's ice nine infecting all the water on Earth and converting it to a new form, the spacetime bubble would continue its conversion of our fragile, 'supercooled' spacetime. A little over four years after the initial disaster, Proxima Centauri, the nearest of our stellar neighbours, would also be annihilated, along with its own planetary system. And then, one by one the other stars of our local group would wink out of existence as the expanding bubble reached them.

One hundred thousand years after that, there would be nothing left of the Milky Way: our entire Galaxy would have been destroyed. Next to go would be the Large and Small Magellanic Clouds, the Milky Way's two small satellite galaxies, and a couple of million years later, the great spiral galaxy in Andromeda, our nearest galactic neighbour, would begin to be eaten by the spacetime bubble. Another 100,000 years later, and Andromeda also would be gone.

Over the next few billion years, galaxy after galaxy would succumb to the relentless spacetime annihilation, until the new spacetime had filled the observable Universe.

Is it really possible that this ultimate disaster could occur? Could human beings one day be to blame for, in effect, blowing up the Universe? In 1983, Martin Rees was visiting the Institute for Advanced Study at Princeton, where he discussed this possibility with a colleague, Piet Hut. They realised that a reasonably comprehensive answer to the question could be found by looking into space, where such particle collisions happen all the time, with cosmic rays crashing into atomic nuclei 'with even greater violence than could be achieved in any currently feasible experiment'.

Rees and Hut concluded that spacetime is not so fragile as to allow this kind of catastrophe to occur: it cannot be 'ripped open' by any energy experiments we are at present conducting, or even planning to conduct. However, Rees adds, if in the future particle accelerators become a hundred times more powerful than the ones we are using today, the danger may reappear. And also, he adds, the collisions occurring in interstellar space are doing so in an environment 'so rarefied that even if they produced a strangelet, it would be unlikely to encounter a third nucleus, so there would be no chance of a runaway process'.

It's a rather different story in the Earth's atmosphere, which contains countless trillions of particles ...

The physicists reassure us that the chances of such a catastrophe occurring are vanishingly small. That may well be, but what if somewhere out there in the depths of space, in some distant star system, an alien civilisation far in advance of us has gained access to undreamed of amounts of energy, and decides to undertake a dangerous experiment such as we have described? It's possible that this has already occurred, and that somewhere, whether it be in our Galaxy or beyond, someone has made a terrible mistake, and the bubble of new spacetime has already annihilated their world and is now expanding outwards into space.

Perhaps the most frightening thing is that if this has happened, we would never know: since the bubble's boundary expands at the speed of light, it would be impossible for us to detect it until it had made contact with the Earth – and annihilated it an instant later.

Sleep well.

27

THE END OF ALL THINGS

THE ULTIMATE FATE OF THE UNIVERSE

The question of how the Universe will end would not seem to be a particularly pressing one. After all, whatever happens will not happen for billions or perhaps trillions of years, long after our own Sun has died and everything we know or can even conceive of has passed into remotest history. Nevertheless, this is one of the greatest unanswered questions of science, and the two options so far examined by cosmologists can be evocatively summed up in the words of two great poets, Robert Frost and T.S. Eliot.

Frost wrote:

> Some say the world will end in fire,
> Some say in ice.
> From what I've tasted of desire,
> I hold with those who favor fire.

Eliot, on the other hand, wrote:

> This is the way the world ends
> This is the way the world ends
> This is the way the world ends
> Not with a bang but a whimper.

Fire or ice; a bang or a whimper. It's a strange fact (and one that is not lost on most cosmologists) that great literature can express in a few lines the very essence of the most intractable and complex of scientific mysteries. At present, cosmologists are faced with two options for the ultimate fate of the Universe; and these options are based upon what they believe to have happened at the very beginning of space and time.

As we saw in the first chapter of this book, it is widely accepted by cosmologists that the universe began approximately 15 billion years ago in the so-called 'Big Bang', in which an infinitesimally small region, containing all the matter and energy we currently observe in the cosmos, exploded in an unthinkably powerful fireball. Such (probably) was the Universe's birth, but what of its ultimate fate? Will the countless billions of stars and galaxies continue to fly apart for ever, embedded in an unstoppably expanding spacetime? Or will the expansion eventually slow to a halt, and then become a contraction, which will culminate after further aeons in an exact mirror image of the creation event, the so-called 'Big Crunch', in which all matter and energy will either be crushed out of existence or 'rebound' into a new universe?

Cosmologists have attempted to discover whether the universal expansion is slowing down, and if so by how much. In 1995 Brian Schmidt of the Mount Stromlo Observatory in Australia began a project to measure the Universe's 'deceleration parameter'. The method which Schmidt and his colleagues used was, first, to look at the nearby Universe and measure how fast it is expanding, and then compare the data with the more distant reaches of the Universe, and compare the two sets of figures. The technique used to make these measurements involved examining the light from Type 1a supernovae, which are bright enough to be seen from across the cosmos, and are uniform enough to allow the accurate calculation of their distance from Earth.

The same project was begun independently by Saul Perlmutter of the Lawrence Berkeley Laboratory in California. Each team of cosmologists realised that what they were expecting to see (the expansion of the Universe slowing down to a greater or lesser degree) was not actually happening. The supernovae which they had expected to be brighter were actually dimmer, implying that the expansion was not slowing down at all, but speeding up.

The Australian and American teams compared results (each had thought that they must have somehow been mistaken in their observations or calculations), and they realised that there was no mistake: the universal expansion really was accelerating, implying the existence of some kind of incredible 'antigravity' effect operating at vast intergalactic scales.

If this accelerated expansion continues (and there doesn't seem to be any reason to suppose it won't), then the ultimate future of the Universe would appear to be cold, dark, boring and very, very dead. Over the next few tens of billions of years, the galaxies we can at present observe with our telescopes will recede further and further into the transgalactic night, until, one by one, they disappear from view. Our Galaxy, the Milky Way, will drift in absolute darkness and solitude through the void.

This will be very long after our Sun has reached the end of its life, swollen into a red giant and then contracted into a relatively tiny white dwarf. Along with Mercury and Venus, Earth would have been swallowed and incinerated in the outer layers of the Sun during its red-giant phase. Many of the other stars will end their lives in spectacular supernova explosions, as the trillions of years pass and the universal expansion continues to increase.

And then, when the last star has exploded or otherwise expired, there will be no light left, anywhere. Where now we see the wondrous profusion of stars and galaxies and nebulae, there will be only the cold, dead husks of planets, the burned-out cinders of dark stars and the silent, open maws of black holes. Eventually, after further billions of years, the black holes will consume those dead stars and planets and the dust drifting between them, until the black holes will be all that remain.

But even that will not quite be the end. Over the next one trillion trillion trillion trillion trillion trillion years, the black holes will themselves evaporate until the Universe is composed of nothing but wandering fundamental particles, forming 'atoms', each of which will be larger than the present-day Universe. And then ... *finally* ... even they will decay, and the Universe will be an indescribably vast, empty void containing absolutely nothing.

This is a profoundly depressing view of the Universe's ultimate destiny; and it is only a theory. It could be that one day in the inconceivably distant future this 'antigravity' force might reverse itself, turning back the clock of eternal expansion. Perhaps . . . perhaps not.

And yet, there is something awe-inspiring in the thought of the ultimate Universe being utterly cold, dark and empty, a limitless, featureless void without thought or movement or physical processes of any kind. One is reminded of the Buddhist concept of Nirvana. Far from being frightening and depressing, might not the final destiny of the Universe be to achieve the perfection of non-existence?

SELECT BIBLIOGRAPHY

Anon., 'The Bennington Triangle', article at www.virtualvermonter.com/
  almanac/benntriangle.htm
Anon., 'Sudden Death of Dinosaurs Questioned', article at http://masseynews.
  massey.ac.nz/2004/Massey_News/oct/oct11/stories/08-18-04.html
Beckley, Timothy Green, *The Smoky God and Other Inner Earth Mysteries*
  (New Brunswick, New Jersey: Inner Light Publications, 1993)
Bennett, Colin, 'Rocket in His Pocket', article on Jack Parsons in *Fortean
  Times* 132
Bord, Janet and Colin, *Alien Animals* (London: Book Club Associates, 1980)
—*Modern Mysteries of the World* (London: Guild Publishing, 1989)
Britt, Robert Roy, 'Brane-Storm Challenges Part of Big Bang Theory', article
  at www.space.com
Brookesmith, Peter, (ed.), *Appearances and Disappearances* (London: Orbis
  Publishing Ltd, 1984)
Carter, John, *Sex and Rockets: The Occult World of Jack Parsons* (Venice,
  California: Feral House, 1999)
Chambers, Paul, 'The Vanishing', article in *Fortean Times* 194
Chapman, Douglas, 'Jack Parsons: Sorcerous Scientist', *Strange Magazine*
  No. 6
Chown, Marcus, *The Universe Next Door* (London: Headline Book
  Publishing, 2002)
Clark, Jerome, *The UFO Book: Encyclopedia of the Extraterrestrial* (Detroit,
  Michigan: Visible Ink Press, 1998)
Cohen, Daniel, *The Encyclopaedia of Monsters* (Waltham Abbey: Fraser
  Stewart, 1982)
Craft, Michael, *Alien Impact* (New York: St Martin's Press, 1996)
Crick, Francis, *The Astonishing Hypothesis: The Scientific Search for the Soul*
  (London: Simon & Schuster Ltd, 1995)
Dash, Mike, 'The Vanishing Lighthousemen of Eilean Mór', essay at www.
  mikedash.com
'Dinosaur Extinction Page', at http://web.ukonline.co.uk/a.buckley/dino.htm
Drake, Frank and Sobel, Dava, *Is Anyone Out There?* (London: Pocket
  Books, 1993)
Farson, Daniel and Hall, Angus, *Mysterious Monsters* (London: Aldus Books,
  1975)

Fort, Charles, *Book of the Damned* (London: John Brown Publishing, 1995)

Fuller, John G., *The Interrupted Journey* (New York: MJF Books, 1996)

Garrett, Richard, *Flight into Mystery: Reports From the Dark Side of the Sky* (London: Weidenfeld & Nicolson, 1986)

Good, Timothy, *Beyond Top Secret: The Worldwide UFO Security Threat* (London: Sidgwick & Jackson, 1996)

Greene, Brian, *The Elegant Universe* (London: Vintage, 2000)

Gribbin, John, *In Search of Schrödinger's Cat* (London: Black Swan, 1993)

—*Companion to the Cosmos* (London: Weidenfeld & Nicolson, 1996)

Guiley, Rosemary Ellen, *Harper's Encyclopedia of Mystical & Paranormal Experience* (Edison, New Jersey: Castle Books, 1991)

Harpur, Patrick, *Daimonic Reality: A Field Guide to the Otherworld* (London: Arkana, 1995)

Hawking, Stephen, *A Brief History of Time* (London: Bantam Press, 1988)

Heidmann, Jean, *Extraterrestrial Intelligence* (Cambridge: Cambridge University Press, 1995)

Horgan, John, *The End of Science* (London: Little, Brown, 1996)

Hoyle, Fred and Wickramasinghe, Chandra, *Our Place in the Cosmos* (London: J.M. Dent Ltd, 1993)

Innes, Brian, *The Catalogue of Ghost Sightings* (London: Blandford Books, 1996)

Keel, John A., *UFOs: Operation Trojan Horse* (London: Abacus, 1973)

—*The Mothman Prophecies* (Avondale Estates, Georgia: IllumiNet Press, 1991)

—*Disneyland of the Gods* (Avondale Estates, Georgia: IllumiNet Press, 1995)

Keith, Jim, *Casebook on the Men in Black* (Avondale Estates, Georgia: IllumiNet Press, 1997)

Keys, David, 'Indonesia's Lost World: Shaking Up the Family Tree', article in *Archaeology*, online publication of the Archaeological Institute of America, at www.archaeology.org/online/features/flores/

Klyce, Brig, 'More Than Panspermia', article at www.panspermia.org/intro.htm

Leakey, Richard, *The Origin of Humankind* (London: Weidenfeld & Nicolson, 1994)

Marcus Aurelius, *Meditations*, tr. Gregory Hays (London: Weidenfeld & Nicolson, 2003)

Marrs, Jim, *Alien Agenda: The Untold Story of the Extraterrestrials Among Us* (London: HarperCollins, 1997)

Mayell, Hillary, 'Comets May Have Led to Birth and Death of Dinosaur Era', article at http://news.nationalgeographic.com/news/2002/05/0516_020516_dinocomet.html

Nichols, Preston B. and Moon, Peter, *The Montauk Project* (Westbury, New York: Sky Books, 1992)

—*Montauk Revisited* (Westbury, New York: Sky Books, 1994)

O'Neill, Terry (ed.), *Out of Time and Place* (St Paul, Minnesota: Llewellyn Publications, 1999)

Peebles, Curtis, *Watch the Skies! A Chronicle of the Flying Saucer Myth* (New York: Berkley Books, 1994)

Pringle, David (ed.), *The Ultimate Encyclopaedia of Science Fiction* (London: Carlton Books, 1996)

Quasar, Gian J., *Into the Bermuda Triangle: Pursuing the Truth Behind the World's Greatest Mystery* (New York: International Marine/McGraw Hill, 2004)

Rees, Martin, *Before the Beginning: Our Universe and Others* (London: Simon & Schuster Ltd, 1997)

—*Our Final Century* (London: William Heinemann, 2003)

Ribeiro, Marcelo B., 'Warp Drive Theory', article at http://omnis.if.ufrj.br/~mbr/warp/

Rickard, Bob and Michell, John, *The Rough Guide to Unexplained Phenomena* (London: Rough Guides Ltd, 2000)

Russell, Davy, 'The Bennington Triangle', article at xprojectmagazine.com

Schoch, Robert M., 'An Enigmatic Ancient Underwater Structure Off the Coast of Yonaguni Island, Japan'; published in Spanish as '*La Pirámide de Yonaguni: ¿Recuerdo de Mu?*', *Más Allá de la Ciencia*, No. 123 (May 1999); available in English at www.robertschoch.net

Steiger, Brad, *The Unknown* (New York: Popular Library, 1966)

—*Mysteries of Time and Space (*New York: Dell Publishing, 1974)

—*Out of the Dark* (New York: Kensington Books, 2001)

Strieber, Whitley, *Communion: A True Story* (New York: William Morrow & Co., 1987)

Sutherly, Curt, *Strange Encounters* (St Paul, Minnesota: Llewellyn Publications, 1996)

Tipler, Frank, *The Physics of Immortality: Modern Cosmology, God and the Resurrection of the Dead* (London: Macmillan, 1995)

Vallée, Jacques, *Dimensions: A Casebook of Alien Contact* (London: Sphere Books, 1990)

—*Revelations: Alien Contact and Human Deception* (New York: Ballantine Books, 1991)

Washington, Peter, *Madame Blavatsky's Baboon: Theosophy and the Emergence of the Western Guru* (London: Secker & Warburg, 1996)

Whitehouse, David, '"Cells" Hint at Life's Origin', article at http://news.bbc.co.uk/1/hi/sci/tech/1142840.stm

—'Faster Than a Speeding Light Wave', article at http://news.bbc.co.uk/1/hi/sci/tech/781199.stm

Wilson, Colin and Wilson, Damon, *The Mammoth Encyclopaedia of Unsolved Mysteries* (London: Constable & Robinson Ltd, 2000)

Wilson, Don, *Our Mysterious Spaceship Moon* (London: Sphere Books, 1976)

Wilson, Robert Anton, *Cosmic Trigger: Final Secret of the Illuminati* (Phoenix, Arizona: New Falcon Publications, 1993)